VIDEO CODING WITH SUPERIMPOSED MOTION-COMPENSATED SIGNALS
Applications to H.264 and Beyond

THE KLUWER INTERNATIONAL SERIES IN ENGINEERING AND COMPUTER SCIENCE

VIDEO CODING WITH SUPERIMPOSED MOTION-COMPENSATED SIGNALS
Applications to H.264 and Beyond

by

Markus Flierl
Signal Processing Institute
Swiss Federal Institute of Technology, Switzerland

Bernd Girod
Information Systems Laboratory
Stanford University, U.S.A.

KLUWER ACADEMIC PUBLISHERS
Boston / Dordrecht / London

Distributors for North, Central and South America:
Kluwer Academic Publishers
101 Philip Drive
Assinippi Park
Norwell, Massachusetts 02061 USA
Telephone (781) 871-6600
Fax (781) 871-6528
E-Mail <kluwer@wkap.com>

Distributors for all other countries:
Kluwer Academic Publishers Group
Post Office Box 322
3300 AH Dordrecht, THE NETHERLANDS
Telephone 31 78 6576 000
Fax 31 78 6576 474
E-Mail <orderdept@wkap.nl>

 Electronic Services <http://www.wkap.nl>

Library of Congress Cataloging-in-Publication

VIDEO CODING WITH SUPERIMPOSED MOTION-COMPENSATED
SIGNALS: *Applications to H.264 and Beyond*
by
Markus Flierl and Bernd Girod
ISBN: 1-4020-7759-9
E-Book ISBN: 1-4020-7765-3

Contents

Preface

This book aims to capture recent advances in motion compensation for efficient video compression. It investigates linearly combined motion compensated signals and generalizes the well known superposition for bidirectional prediction in B-pictures. The number of superimposed signals and the selection of reference pictures will be important aspects of the discussion.

The application oriented part of the book employs this concept to the well known ITU-T Recommendation H.263 and continues with the improvements by superimposed motion-compensated signals for the emerging ITU-T Recommendation H.264 and ISO/IEC MPEG-4 (Part 10). In addition, it discusses a new approach for wavelet-based video coding. This technology is currently investigated by MPEG to develop a new video compression standard for the mid-term future.

The theoretical part of the book provides a deeper understanding of the underlying principles of superimposed motion-compensated signals. The text incorporates more than 200 references, summarizes relevant prior work, and develops a mathematical characterization of superimposed motion-compensated signals. The derived information-theoretic performance bounds permit a valuable comparison of the investigated compression schemes.

Acknowledgments

This work required a substantial effort and I am very grateful to many people who made this endeavor possible. I would like to thank Prof. Bernd Girod for the opportunity to join his group in Erlangen and Stanford as well as to benefit from the inspiring environment. I thank Prof. André Kaup, Prof. Heinrich Niemann, and Prof. Wolfgang Koch for their interest in this work and, in particular, Dr. Nick Kingsbury for helpful discussions. I am thankful to Anne Aaron, Chuo-Ling Chang, Jacob Charareski, Joachim Eggers, Peter Eisert, Niko Färber, Sang-Eun Han, Frank Hartung, Mark Kalman, Yi Liang, Marcus Magnor, James Mammen, Prashant Ramanathan, Shantanu Rane, Marion Schabert, Eric Setton, Wolfgang Sörgel, Eckehard Steinbach, Klaus Stuhlmüller, Jonathan Su, Matthias Teschner, Lutz Trautmann, Thomas Wiegand, and Rui Zhang for many stimulating discussions, joint work, and proofreading. I am also grateful to the people at the Telecommunications Laboratory and Information Systems Laboratory, especially to Ursula Arnold and Kelly Yilmaz for their invaluable administrative support. My special thanks belong to my family and friends for all the support they gave.

Markus Flierl

Chapter 1

INTRODUCTION

Motion-compensated prediction is an important component of current hybrid video coding systems. In recent years, advances in compression efficiency have mainly been achieved by improved motion-compensated prediction, e.g. sub-pel accurate motion compensation [1], variable block size prediction [2], multiframe prediction [3–5], and multihypothesis motion compensation.

Multihypothesis motion-compensated prediction linearly combines multiple motion-compensated signals to arrive at the actual prediction signal. This term was first used in [6] to provide a framework for overlapped block motion compensation (OBMC). OBMC was introduced to reduce blocking artifacts in motion-compensated prediction [7]. In earlier work, attempts have been made to combine signals from more than one frame. Published in 1985, [8] investigates adaptive predictors for hybrid coding that use up to four previous fields. In the same year, the efficiency of bidirectional prediction has been raised in [9]. To predict the current frame, bidirectional prediction uses a linear combination of two motion-compensated signals: one is chosen from the next reference frame, the other from the previous reference frame. Bidirectional prediction characterizes the now known concept of B-pictures which has originally been proposed to MPEG [10]. The motivation was to interpolate any skipped frame taking into account the movement between the two "end" frames. The technique, originally called conditional motion-compensated interpolation, coupled the motion-compensated interpolation strategy with transmission of significant interpolation errors.

These practical schemes have been studied in [11] and summarized in a theoretical analysis of multihypothesis motion-compensated prediction. The analysis is based on a power spectral density model for inaccurate motion compensation [12] which has been proven successful to characterize motion-compensated prediction. Variations of these practical schemes have also been standardized in, e.g., [13] and [14].

With the advent of multiframe prediction, the question of linearly combined prediction signals has been re-visited. [15, 16] design general predictors for block-based superimposed motion compensation that utilize several previous reference frames. To determine an efficient set of motion-compensated blocks, an iterative algorithm is used to improve successively conditional optimal motion-compensated blocks. A similar algorithm has been proposed in [17] for bidirectional prediction only.

Multiple frames are not only used for predictive coding. Schemes for three-dimensional subband coding of video signals consider also multiple frames [18, 19]. Adaptive wavelet transforms with motion compensation can be used for temporal subband decomposition [20]. These schemes use again linear combinations of motion-compensated signals and are also of interest for our investigations.

This book "Video Coding with Superimposed Motion-Compensated Signals", contributes to the field of motion-compensated video coding as follows:

1 For video compression, we investigate the efficiency of block-based superimposed prediction with multiframe motion compensation based on the ITU-T Rec. H.263. We explore the efficient number of superimposed prediction signals, the impact of variable block sizes compensation, and the influence of the size of the multiple reference frame buffer if the reference frames are previous frames only.

2 We generalize B-pictures for the emerging ITU-T Rec. H.264 to the generic concept of superimposed prediction which chooses motion-compensated blocks from an arbitrary set of reference pictures and measure the improvement in compression efficiency. Further, the generic concept of superimposed prediction allows also that generalized B-pictures are used for reference to predict other B-pictures. As this is not the case for classic B-pictures, we explore the efficiency of this aspect too.

3 We build on the theory of multihypothesis motion-compensated prediction for video coding and extend it to motion-compensated prediction with complementary hypotheses. We assume that the displacement errors of the multiple hypotheses are jointly distributed and, in particular, correlated. We investigate the efficiency of superimposed motion compensation as a function of the displacement error correlation coefficient. We conclude that

compensation with complementary hypotheses results in maximally negatively correlated displacement error. We continue and determine a high-rate
approximation for the rate difference with respect to optimal intra-frame
encoding and compare the results to [11]. To capture the influence of multiframe motion compensation, we model it by forward-adaptive hypothesis
switching and show that switching among M hypotheses with statistically
independent displacement error reduces the displacement error variance by
up to a factor of M.

4 We do not only consider predictive coding with motion compensation. We
   explore also the combination of linear transforms and motion compensation for a temporal subband decomposition of video and discuss motion
   compensation for groups of pictures. Therefore, we investigate experimentally and theoretically motion-compensated lifted wavelets for three-
   dimensional subband coding of video. The experiments capture the coding
   efficiency dependent on the number of pictures in the group and permit
   a comparison to predictive coding with motion compensation. The theoretical discussion analyzes the investigated lifted wavelets and builds on
   the insights from motion compensation with multiple hypotheses. Further,
   the analysis provides performance bounds for three-dimensional subband
   coding with motion compensation and gives insight about potential coding
   gains.

This book is organized as follows: Chapter 2 provides the background for
video coding with superimposed motion-compensated signals and discusses
related work. Chapter 3 investigates motion-compensated prediction with
complementary hypotheses. Based on ITU-T Rec. H.263, Chapter 4 explores
experimental results for video coding with superimposed motion-compensated
prediction and multiple reference frames. Chapter 5 discusses generalized B-
pictures for the emerging ITU-T Rec. H.264. Finally, Chapter 6 explores linear
transforms with motion compensation and its application to motion compensation for groups of pictures.

Chapter 2

BACKGROUND AND RELATED WORK

2.1 Coding of Video Signals

Standard video codecs like ITU-T Recommendation H.263 [14, 21] or the emerging ITU-T Recommendation H.264 [22] are hybrid video codecs. They incorporate an intra-frame codec and a motion-compensated predictor. The intra-frame codec is able to encode and decode one frame independently of others, whereas the motion-compensated predictor is able to compensate motion between the current frame and a previously decoded frame.

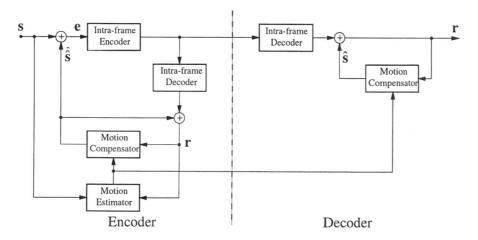

Figure 2.1. Hybrid video codec utilizing motion-compensated prediction.

Fig. 2.1 depicts such a hybrid video codec. The encoder estimates the motion between the current frame **s** and a previously decoded frame **r** and transmits it as side information to the decoder. Both encoder and decoder use motion compensation to generate the motion-compensated frame $\hat{s}$ from previously reconstructed frames which are also available at the decoder. Therefore, only the frame difference $\mathbf{e}$ between the current frame and the motion-compensated frame needs to be encoded by the intra-frame encoder. This frame difference has much less signal energy than the current frame and, hence, requires less bit-rate if encoded. Despite the side information, the overall bit-rate of a hybrid video codec is less than the bit-rate of a video codec with intra-frame coding only. Therefore, motion-compensated prediction is an important component for efficient compression with a hybrid video codec.

Hybrid video codecs require sequential processing of video signals which makes it difficult to achieve efficient embedded representations of video sequences. Therefore, we consider also motion-compensated three-dimensional subband coding of video signals [23–25]. Applying a linear transform in temporal direction of a video sequence may not be very efficient if significant motion is prevalent. Motion compensation between two frames is necessary to deal with the motion in a sequence. A combination of linear transform and motion compensation is required for efficient three-dimensional subband coding.

In the following, we review relevant techniques and principles for state-of-the-art video coding. The discussion provides a background for the following chapters and summarizes work on which we will build. Section 2.2 outlines several relevant methods for motion-compensation: bidirectional motion compensation, overlapped block motion compensation, variable block size motion compensation, multiframe motion compensation, and superimposed motion compensation. Section 2.3 discusses previous work on rate-constrained motion estimation, rate-constrained estimation of superimposed motion, quantizer selection at the residual encoder, and techniques for efficient motion estimation. Section 2.4 introduces to a theory for motion-compensated prediction. We discuss the underlying frame signal model, review the model for motion-compensated prediction, and outline the state-of-the-art for multihypothesis motion-compensated prediction. We reference the utilized performance measures and summarize the results of this theory. Finally, Section 2.5 summarizes previous work on three-dimensional subband coding of video. We outline the problem of motion compensation for the temporal subband decomposition and refer to adaptive lifting schemes that permit motion compensation.

2.2 Motion Compensation

The efficiency of inter-frame coding schemes for video sequences is improved by motion compensation. Efficient motion compensation requires an accurate measurement of the displacement field between two frames. A practical algorithm for this measurement is block matching [26] [27]. It estimates the displacement field on a block bases, i.e., approximates each block with one displacement value by minimizing its prediction error. Efficient motion-compensated coding is desirable especially for low bit-rate video applications [28]. Usually, block-based schemes assume just translatory motion for all pels in the block. But more sophisticated schemes like spatial transforms [29] and transformed block-based motion compensation [30] are possible. And by omitting the block constraint, arbitrarily shaped regions can be used for motion compensation [31].

Efficient inter-frame coding schemes consider also the problem of noise reduction in image sequences. DUBOIS and SABRI describe in [32] a nonlinear temporal filtering algorithm using motion compensation. WOODS et al. present a spatio-temporal adaptive 3-D Kalman filter with motion compensation [33] and couple it with motion estimation in [34].

2.2.1 Bidirectional Motion Compensation

Frame skipping is a viable technique to reduce drastically the bit-rate necessary to transmit a video signal. If all frames have to be reconstructed at the decoder, skipped frames must be interpolated by a motion compensation scheme. NETRAVALI and ROBBINS propose such a scheme in [35] and initiate further research in this field. SORYANI and CLARKE combine image segmentation and adaptive frame interpolation [36], THOMA and BIERLING consider covered and uncovered background for motion-compensated interpolation [37], and CAFFORIO et al. discuss a pel-recursive algorithm [38]. Similar algorithms for adaptive frame interpolation are outlined in [39–42]. KOVAČEVIĆ et al. use adaptive bidirectional interpolation for deinterlacing [43, 44]. AU et al. study also temporal frame interpolation [45] and compare block- and pixel-based interpolation in [46]. They propose temporal interpolation with overlapping [47] and unidirectional motion-compensated temporal interpolation [48]. They also suggest zonal based algorithms for temporal interpolation [49]. These algorithms allow efficient block-based motion estimation [50, 51].

To improve the quality of the interpolated frames at the decoder, the interpolation error is encoded and additional bits are transmitted to the decoder. MICKE considers first the idea of interpolation error encoding in [52]. HASKELL and PURI [53] as well as YONEMITSU and ANDREWS [54] propose algorithms for this approach. This type of coded picture is also called B-picture. PURI et al. investigate several aspects of this picture type, like quantization and temporal resolution scalability [55–58]. Additionally, SHANABLEH and GHANBARI point out the importance of B-pictures in video streaming [59].

LYNCH shows how to derive B-picture motion vectors from neighboring P-pictures [60]. But the displacement field of the B-pictures can also be encoded and transmitted to the decoder. For example, WOODS et al. suggest compactly encoded optical-flow fields and label fields for frame interpolation and bidirectional prediction [61].

B-pictures, as they are employed in MPEG-1 [62], MPEG-2 [13], or H.263 [63], utilize *Bidirectional Motion Compensation.* Bidirectional motion-compensated prediction is an example for multihypothesis motion-compensated prediction where two motion-compensated signals are superimposed to reduce the bit-rate of a video codec. But the concept of B-pictures has to deal with a significant drawback: prediction uses the reference pictures before and after the B-pictures as depicted in Fig. 2.2. The associated delay of several frames may be unacceptable for interactive applications. In addition, the two motion vectors always point to the previous and subsequent frames and the advantage of a variable picture reference cannot be exploited.

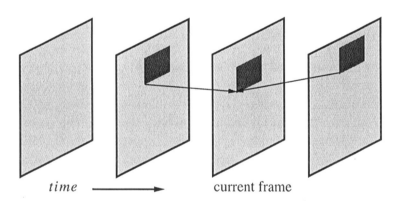

time ⟶   current frame

Figure 2.2.   Bidirectional motion compensation. Reference pictures are used only before and after the current frame.

The efficiency of forward and backward prediction has already been raised by MUSMANN et al. [9] in 1985. The now known concept of B-pictures was proposed to MPEG by PURI et al. [10]. The motivation was to interpolate any skipped frame taking into account the movement between the two "end" frames. The technique, called conditional motion-compensated interpolation, couples the motion-compensated interpolation strategy with transmission of the significant interpolation errors.

For joint estimation of forward and backward motion vectors in bidirectional prediction, a low-complexity iterative algorithm is introduced in [64]. Starting from the initial values obtained by a commonly-used block matching independent search method, the motion vectors are iteratively refined until a locally optimal solution to the motion estimation problem for bidirectional prediction is achieved. Each iteration consists of a series of two similar procedures. First, the backward motion vector is fixed and a new forward motion vector is searched to minimize the prediction error. Then the forward motion vector is fixed and the backward motion vector is similarly refined by minimizing the prediction error. This process is repeated until the prediction error no further decreases. The iterative search procedure minimizes the prediction error and considers no rate constraint [17]. The price paid for the improvement in performance is only a small increase in computational complexity relative to independent search for the two motion vectors. Experiments in [17] show that the increase in search complexity is, on average, less than 20% of that of the independent search. Based on this work, [65] proposes an efficient motion estimation algorithm.

2.2.2 Overlapped Block Motion Compensation

Like bidirectional prediction, *Overlapped Block Motion Compensation* (OBMC) [7, 6, 66–68] is another example of the general concept of multi-hypothesis motion-compensated prediction. Originally, the motivation was to reduce blocking artifacts caused by block motion compensation. SULLIVAN and BAKER introduced motion compensation using control grid interpolation [69] and WATANABE and SINGHAL windowed motion compensation [70]. OBMC uses more than one motion vector for predicting the same pixel but, in contrast to bidirectional prediction, does not increase the number of vectors per block.

The discussion of overlapped compensation in [6] is based on 'multi-hypothesis expectation'. The paper argues that only one block motion vector is encoded for a large block of pixels and that the vector value is limited in precision such that an encoded block motion vector may not be correct for all pixels in the block. For this reason, [6] proposes a multi-hypothesis expectation paradigm. Since the decoder cannot know the correct motion vector for each pixel, the motion uncertainty is modeled by a posteriori inferred displacement probability density function, conditioned on the encoded data. Using this distribution, an ideal decoder could generate a minimum mean square error estimate for each pixel prediction. A vector of linear weights and an associated set of displacements are defined to determine the prediction for each pixel. It is reported that the method effectively eliminates blocking artifacts and reduces prediction error.

[67] proposes an estimation-theoretic paradigm for analyzing and optimizing the performance of block-based motion compensation algorithms. OBMC is derived as a linear estimator of each pixel intensity, given that the only

motion information available to the decoder is a set of block-based vectors. OBMC predicts the frame of a sequence by repositioning overlapping blocks of pixels from the reference frame, each weighted by some smooth window. The estimation-theoretic formulation leads directly to statistical techniques for optimized window design. The work also considers the problem of optimized motion estimation. Viewed within the estimation-theoretic paradigm for block-based motion compensation, the objective of motion estimation is to provide the decoder information that optimizes the performance of its prediction. Optimal motion estimation for OBMC involves estimating a noncausally related motion field, and an iterative procedure is proposed for solving this problem.

Tao and Orchard continue the investigation. With the goal to remove motion uncertainty and quantization noise [71], they discuss OBMC and loop filtering in [72], propose a method for window design [73], and investigate noniterative motion estimation [74]. They also suggest a gradient-based model for the residual variance [75] and propose a parametric solution for OBMC [76].

2.2.3 Variable Block Size Motion Compensation

Motion-compensated prediction with blocks of variable size improves the efficiency of video compression algorithms by adapting spatially displacement information [77, 2, 78–82]. *Variable Block Size* (VBS) prediction assigns more than one motion vector per macroblock but it uses just one motion vector for a particular pixel.

[2] describes a method for optimizing the performance of block-based motion-compensated prediction for video compression using fixed or variable size blocks. A Lagrangian cost function is used to choose motion vectors and block sizes for each region of the prediction image, that gives the best performance in a rate-distortion sense. For that, a quadtree is used to structure blocks of variable size. The variable block size algorithm determines the optimal tree structure and yields a significant improvement in rate-distortion performance over motion compensation with a fixed block size.

[81, 82] investigate a more general tree structure for motion- and intensity-compensated video coding. In contrast to variable block size motion compensation, this approach incorporates also the intensity residual into the tree structure. The work discusses pruning and growing algorithms to determine rate-distortion optimal tree structures by utilizing a Lagrangian cost function.

ITU-T Rec. H.263 [14] provides variable block size capability. The INTER4V coding mode allows 8×8 blocks in addition to the standard 16×16 blocks. VBS motion-compensated prediction utilizes either OBMC or an in-loop deblocking filter to reduce blocking artifacts. The emerging ITU-T Rec. H.264 [22] allows up to 7 different block sizes from 16×16 to 4×4. Here, the blocking artifacts are reduced by an in-loop deblocking filter.

2.2.4   Multiframe Motion Compensation

Multiframe techniques have first been utilized for background prediction by MUKAWA and KURODA [83]. The method permits a prediction signal for uncovered background. With a special background memory, the method is also investigated by HEPPER [84]. LAVAGETTO and LEONARDI discuss block-adaptive quantization of multiple-frame motion fields [85]. GOTHE and VAISEY improve motion compensation by using multiple temporal frames [3]. They provide experimental results for 8×8 block-motion compensation with up to 8 previous frames. FUKUHARA, ASAI, and MURAKAMI propose low bit-rate video coding with block partitioning and adaptive selection of two time-differential frame memories [86]. The concept of *multiframe motion-compensated prediction* is used by BUDAGAVI and GIBSON to control error propagation for video transmission over wireless channels [4, 87–89].

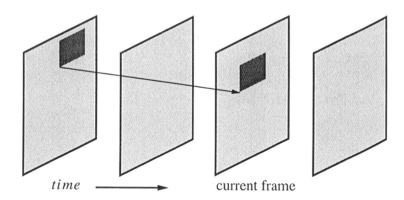

time ⟶   current frame

Figure 2.3.   Multiframe motion compensation. A reference frame is chosen by an additional picture reference parameter.

Long-term memory motion-compensated prediction [5, 90–93] employs several reference frames, i.e., several previously decoded frames, whereas standard motion-compensated prediction utilizes one reference frame, i.e. the previously decoded frame. This is accomplished by assigning a variable picture reference parameter to each block motion vector as shown in Fig. 2.3. The additional reference parameter overcomes the restriction that a specific block has to be chosen from a certain reference frame. This generalization improves compression efficiency of motion-compensated prediction.

In [91], the long-term memory covers several seconds of decoded frames at the encoder and decoder. The use of multiple frames for motion compensation in most cases provides significantly improved prediction gain. The variable picture reference that permits the use of several frames has to be transmitted as side information requiring an additional bit-rate which may be prohibitive when the size of the long-term memory becomes large. Therefore, the bit-rate of the motion information is controlled by employing rate-constrained motion estimation to trade-off the better prediction signal against the increased bit-rate.

Multiframe block motion compensation in [89] makes use of the redundancy that exists across multiple frames in typical video conferencing sequences to achieve additional compression over that obtained by using single frame block motion compensation. The multiframe approach also has an inherent ability to overcome some transmission errors and is thus more robust when compared to the single frame approach. Additional robustness is achieved by randomized frame selection among the multiple previous frames.

Annex U of ITU-T Rec. H.263 [14], entitled "Enhanced Reference Picture Selection Mode," provides multiframe capability for both improved compression efficiency and enhanced resilience to temporal error propagation due to transmission errors. The multiframe concept is also incorporated into the emerging ITU-T Rec. H.264 [22].

2.2.5   Superimposed Motion Compensation

Standard block-based motion compensation approximates each block in the current frame by a spatially displaced block chosen from the previous frame. As an extension, long-term memory motion compensation chooses the blocks from several previously decoded frames [91]. The motion-compensated signal is determined by the transmitted motion vector and picture reference parameter.

Now, consider N motion-compensated signals, also called hypotheses. The superimposed prediction signal is the linear superposition of these N hypotheses. Constant scalar coefficients determine the weight of each hypothesis for the predicted block. Only N scalar coefficients are used and each coefficient is applied to all pixel values of the corresponding hypothesis. That is, spatial filtering and OBMC are not employed. Note, that weighted averaging of a set of images is advantageous in the presence of noise as discussed by UNSER and EDEN in [94].

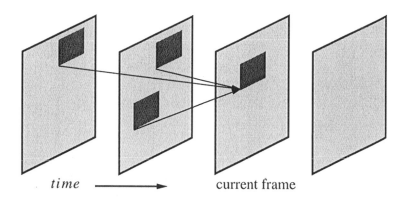

time ⟶ current frame

Figure 2.4. Superimposed motion-compensated prediction with three hypotheses. Three blocks of previous decoded frames are linearly combined to form a prediction signal for the current frame.

Fig. 2.4 shows three hypotheses from previous decoded frames which are linearly combined to form the superimposed prediction signal for the current frame. Please note that a hypothesis can be chosen from any reference frame. Therefore, each hypothesis has to be assigned an individual picture reference parameter [15, 16, 95, 96].

This scheme differs from the concept of B-frame prediction in three significant ways: First, all reference frames are chosen from the past. No reference is made to a subsequent frame, as with B-frames, and hence no extra delay is incurred. Second, hypotheses are not restricted to stem from particular reference frames due to the picture reference parameter. This enables the encoder to find a much more accurate set of prediction signals, at the expense of a minor increase in the number of bits needed to specify them. Third, it is possible to combine more than two motion-compensated signals. As will be shown later, these three properties of superimposed motion compensation improve the coding efficiency of a H.263 codec without incurring the delay that would be caused by using B-pictures.

A special forward prediction mode with two averaged prediction signals is specified in MPEG-2 [13, 97]. This mode is called "Dual Prime Prediction" and can be used for predictive compression of interlaced video. It is utilized in P-pictures where there are no B-pictures between the predicted and the reference pictures. Dual prime is a forward field prediction where a single forward motion vector is estimated for each macroblock of the predicted frame picture. This motion vector points at the reference frame which is the most recent reconstructed frame. Using this vector, each field in the macroblock is associated with a field of the same parity in the reference frame. Two motion vectors pointing at fields of opposite parity are derived from the estimated motion vector by assuming a linear motion trajectory. That is, the motion vectors are scaled according to the temporal distance between the reference and the predicted frames. The two motion-compensated signals are simply averaged to form the prediction.

The design of the superimposed motion-compensated predictor is such that the mean square prediction error is minimized while limiting the bit-rate consumed by the motion vectors and picture reference parameters. With variable length coding of the side information, the best choice of hypotheses will depend on the code tables used, while the best code tables depend on the probabilities of choosing certain motion vector/reference parameter combinations. Further, the best choice of hypotheses also depends on the linear coefficients used to weight each hypothesis, while the best coefficients depend on the covariance matrix of the hypotheses.

To solve this design problem, it is useful to interpret superimposed motion-compensated prediction as a vector quantization problem [98–100]. *Entropy Constrained Vector Quantization* (ECVQ) [101, 102], which is an extension of the *Generalized Lloyd Algorithm* [103], is employed to solve the design problem iteratively. For the interpretation of motion-compensated prediction, we argue that a block in the current frame is quantized. The output index of the quantizer is the index of the displacement vector. Each displacement vector is represented by a unique entropy codeword. Further, the codebook used for quantization contains motion-compensated blocks chosen from previous frames. This codebook is adaptive as the reference frames change with the current frame. For superimposed prediction, the codebook contains N-tuple of motion-compensated blocks whose components are linearly combined.

Rate-constrained superimposed motion estimation utilizes a Lagrangian cost function. The costs are calculated by adding the mean square prediction error to a rate-term for the motion information, which is weighted by a Lagrange multiplier [104]. The estimator minimizes this cost function on a block basis to determine multiple displacement parameter. This corresponds to the biased nearest neighbor condition familiar from vector quantization with rate constraint. The decoder combines linearly more than one motion-compensated signal which are determined by multiple displacement parameter. In [15, 16], several video sequences are encoded to show that the converged design algorithm just averages multiple hypotheses.

Superimposed motion compensation requires the estimation of multiple motion vectors and picture reference parameters. Best prediction performance is obtained when the N motion vectors and picture reference parameters are jointly estimated. This joint estimation is computationally very demanding. Complexity can be reduced by an iterative algorithm which improves conditional optimal solutions step by step [15, 16]. More details are discussed in Section 2.3.2.

It can be observed that increasing the number of hypotheses also improves the quality of the prediction signal. The gains by superimposed prediction with multiframe motion compensation are larger than with single frame motion compensation. That is, superimposed motion-compensated prediction benefits from multiframe motion compensation such that the PSNR prediction gain is more than additive [15, 16].

It is important to note that an N-hypothesis uses N motion vectors and picture reference parameters to form the prediction signal. Applying a product code for these N references will approximately increase the bit-rate for N-hypothesis MCP by factor of N. This higher rate has to be justified by the improved prediction quality.

When predicting a target block in the current frame, we have the choice of several different predictors, i.e. 1-hypothesis, 2-hypothesis, ... N-hypothesis predictor. Experiments reveal that each predictor on its own is not the best one in the rate-distortion sense. For the same prediction quality, the 1-hypothesis predictor provides always the lowest bit-rate. On the other hand, improved prediction quality can only be obtained by increasing the number of hypotheses. Therefore, the optimal rate-distortion performance results from selecting the predictor that gives the best rate-distortion performance. Moreover, this selection depends on the block to be predicted [15, 16].

2.3   Motion Estimation

Classic motion estimation aims to minimize the energy in the displaced frame difference. But actually, it is a bit allocation problem for both motion vectors and displaced frame difference as they are dependent. RAMCHANDRAN, ORTEGA and VETTERLI investigate in [105] the problem of bit allocation for dependent quantization. They apply it to the problem of frame type selection like LEE and DICKINSON in [106]. But RIBAS-CORBERA and NEUHOFF in [107] as well as SCHUSTER and KATSAGGELOS in [108] discuss in detail bit allocation between displacement and displaced frame difference. Additional segmentation of the displacement field is investigated in [109] and [110]. WOODS et al. discuss motion vector quantization for video coding [111, 112] as well as multiscale modeling and estimation of motion fields [113]. *Rate-constrained motion estimation* which utilizes a Lagrangian cost function is discussed by several authors [114–118].

2.3.1   Rate-Constrained Motion Estimation

In the following discussion, we relate the problem of block-based motion-compensated prediction to vector quantization with a rate constraint [15]. For block-based motion-compensated prediction, each block in the current frame is approximated by a spatially displaced block chosen from a reference picture. Each $a \times a$ block is associated with a vector in the a^2-dimensional space $\mathcal{R}^{a^2}$. A original block is represented by the vector-valued random variable **s**. A particular original block is denoted by s. The quality of motion-compensated prediction is measured by the average sum square error distortion between original **s** and predicted blocks **ŝ**.

$$D = E\left\{ \|\mathbf{s} - \hat{\mathbf{s}}\|_2^2 \right\} \tag{2.1}$$

The blocks are coded with a displacement code **b**. Each displacement codeword b provides a unique rule how to approximate the current block-sample s. The average rate of the displacement code is determined by its average length.

$$R = E\left\{ |\mathbf{b}| \right\} \tag{2.2}$$

Rate-distortion optimal motion-compensated prediction minimizes average prediction distortion subject to a given average displacement rate. This constrained problem can be converted to an unconstrained problem by defining a Lagrangian cost function J with a Lagrange multiplier λ [104, 101, 100, 119].

$$J(\lambda) = E\left\{ \|\mathbf{s} - \hat{\mathbf{s}}\|_2^2 + \lambda\,|\mathbf{b}| \right\} \tag{2.3}$$

The predictor with the minimum Lagrangian costs is also a rate-distortion optimal predictor.

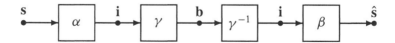

Figure 2.5.   Interpreting motion-compensated prediction as vector quantization.

The ECVQ algorithm [101] suggests a vector quantizer model according to Fig. 2.5. This model is interpreted for motion-compensated prediction as follows: Given the original block **s**, the mapping α estimates the best displacement index **i** in the codebook of reference blocks (frame memory). The mapping γ assigns a variable length codeword to each displacement index. To be lossless, γ has to be invertible and uniquely decodable [101]. For block-based motion-compensated prediction, β is a codebook look-up of reference blocks to determine the block $\hat{s}$ that is used for prediction.

Minimizing the Lagrangian costs (2.3) provides the rate-distortion optimum predictor, that is, the optimum mappings α and γ. For that, the Lagrangian cost function (2.3) is expressed in terms of the model mappings α and γ,

$$J(\alpha, \beta, \gamma, \lambda, s) = E\left\{\|s - \beta \circ \alpha(s)\|_2^2 + \lambda|\gamma \circ \alpha(s)|\right\} \tag{2.4}$$

with the blocks $\hat{s} = \beta \circ \alpha(s)$ used for prediction and the codewords $\mathbf{b} = \gamma \circ \alpha(s)$. For a given distribution of the original blocks s_c in the training set and constant Lagrange multiplier λ_c, the optimal predictor incorporates the optimal mappings α and γ which satisfy

$$\min_{\{\alpha, \gamma\}} J(\alpha, \beta, \gamma, \lambda_c, s_c). \tag{2.5}$$

Given the distribution of the original blocks s_c in the training set as well as the Lagrange multiplier λ_c, an iterative design algorithm for solving (2.5) includes two steps. For motion-compensated prediction, β_c retrieves the compensated block from the frame memory, which is simply a codebook look-up. The first step determines the optimal displacement index $\mathbf{i} = \alpha(s)$ for the given mapping γ_c.

$$\min_{\{\alpha\}} E\left\{\|s_c - \beta_c \circ \alpha(s_c)\|_2^2 + \lambda_c|\gamma_c \circ \alpha(s_c)|\right\}$$
$$\implies \alpha(s) = \operatorname*{argmin}_{\{i\}}\left\{\|s - \beta_c(i)\|_2^2 + \lambda_c|\gamma_c(i)|\right\} \tag{2.6}$$

(2.6) is the biased nearest neighbor condition familiar from vector quantization with a rate-constraint. The second step determines the optimal entropy coding γ for the given motion estimation α_c. For a given α_c, the distribution of the displacement indices $\mathbf{i}_c$ is constant.

$$\min_{\{\gamma\}} E\left\{\|s_c - \beta_c \circ \alpha_c(s_c)\|_2^2 + \lambda_c|\gamma \circ \alpha_c(s_c)|\right\}$$
$$\implies \min_{\{\gamma\}} E\left\{|\gamma(\mathbf{i}_c)|\right\} \tag{2.7}$$

(2.7) postulates the minimum average codeword length of the displacement code, given the displacement indices. This problem can be solved with, e.g., the Huffman algorithm. Finally, given the entropy code γ, the problem of rate-constrained motion estimation is solved by (2.6).

2.3.2   Rate-Constrained Estimation of Superimposed Motion

For rate-constrained estimation of superimposed block motion, the same methodology is employed as outlined for rate-constrained motion estimation. Each block in the current frame is approximated by more than one spatially displaced block chosen from the set of reference pictures. Each $a \times a$ block is associated with a vector in the a^2-dimensional space $\mathcal{R}^{a^2}$. A original block is represented by the vector-valued random variable $\mathbf{s}$. A particular original block is denoted by s.

The vector quantizer model according to Fig. 2.5 is interpreted for motion-compensated prediction with N superimposed blocks as follows: Given the original block $\mathbf{s}$, the mapping α estimates again the best displacement index $\mathbf{i}$. The mapping γ assigns a variable length codeword to each displacement index. γ is invertible and uniquely decodable. For motion-compensated prediction with superimposed blocks, β is a codebook look-up of a row-vector $\mathbf{c} = (\mathbf{c}_1, \mathbf{c}_2, \ldots, \mathbf{c}_N)$ of N motion-compensated blocks. These N motion-compensated blocks are superimposed and determine the block $\hat{\mathbf{s}}$ that is used to predict $\mathbf{s}$. The weighted superposition of N motion-compensated blocks is accomplished with N scalar weights h_μ, $\mu = 1, 2, \ldots, N$, that sum to one. For simplicity, the vector of scalar weights is denoted by h.

$$\hat{\mathbf{s}} = \mathbf{c}h = (\mathbf{c}_1, \mathbf{c}_2, \ldots, \mathbf{c}_N) \begin{pmatrix} h_1 \\ h_2 \\ \vdots \\ h_N \end{pmatrix} \tag{2.8}$$

For the scheme of superimposed motion, minimizing the Lagrangian costs (2.3) provides the rate-distortion optimum predictor for superimposed blocks, that is, the optimum mappings α, β, and γ. The Lagrangian cost function (2.3) is expressed in terms of the model mappings α, β, and γ according to (2.4). The blocks $\hat{\mathbf{s}} = \beta \circ \alpha(\mathbf{s})$ are used to predict $\mathbf{s}$, and the codewords $\mathbf{b} = \gamma \circ \alpha(\mathbf{s})$ to encode the displacement indices. For a given distribution of the original blocks $\mathbf{s}_c$ in the training set and constant Lagrange multiplier λ_c, the optimal predictor incorporates the optimal mappings α, β, and γ which satisfy

$$\min_{\{\alpha, \beta, \gamma\}} J(\alpha, \beta, \gamma, \lambda_c, \mathbf{s}_c). \tag{2.9}$$

Given the distribution of the original blocks $\mathbf{s}_c$ in the training set as well as the Lagrange multiplier λ_c, an iterative design algorithm for solving (2.9) includes three steps. For superimposed motion-compensated prediction, β is also optimized. The first step determines the optimal displacement indices $\mathbf{i} = \alpha(\mathbf{s})$ for the given mappings β_c and γ_c.

$$\min_{\{\alpha\}} E\left\{\|\mathbf{s}_c - \beta_c \circ \alpha(\mathbf{s}_c)\|_2^2 + \lambda_c |\gamma_c \circ \alpha(\mathbf{s}_c)|\right\}$$

$$\implies \alpha(s) = \operatorname*{argmin}_{\{i\}} \left\{\|s - \beta_c(i)\|_2^2 + \lambda_c|\gamma_c(i)|\right\} \quad (2.10)$$

(2.10) is the biased nearest neighbor condition familiar from vector quantization with a rate-constraint. The second step determines the optimal entropy coding γ for the given estimation of superimposed motion α_c. For given α_c, the distribution of the displacement indices $\mathbf{i}_c$ is constant.

$$\min_{\{\gamma\}} E\left\{\|\mathbf{s}_c - \beta_c \circ \alpha_c(\mathbf{s}_c)\|_2^2 + \lambda_c |\gamma \circ \alpha_c(\mathbf{s}_c)|\right\}$$

$$\implies \min_{\{\gamma\}} E\left\{|\gamma(\mathbf{i}_c)|\right\} \quad (2.11)$$

(2.11) postulates a minimum average codeword length of the displacement code, given the displacement indices. The third step determines the optimal superposition, i.e., the optimal scalar weights, given the mappings α_c and γ_c.

$$\min_{\{\beta\}} E\left\{\|\mathbf{s}_c - \beta \circ \alpha_c(\mathbf{s}_c)\|_2^2 + \lambda_c |\gamma_c \circ \alpha_c(\mathbf{s}_c)|\right\}$$

$$\implies \min_{\{h:\mathbf{1}^T h=1\}} E\left\{\|\mathbf{s}_c - \mathbf{c}_c h\|_2^2\right\} \quad (2.12)$$

(2.12) is the Wiener problem for the conditional optimal superposition coefficients. The superimposed predictor preserves the expected value of the original block, i.e., $E\{s\} = E\{\hat{s}\}$. Consequently, the Wiener problem can be expressed in covariance notation

$$\min_{\{h:\mathbf{1}^T h=1\}} \left\{C_{ss} - 2h^T C_{cs} + h^T C_{cc} h\right\} \quad (2.13)$$

where C_{ss} is the scalar variance of the original blocks, C_{cc} the $N \times N$ covariance matrix of the motion-compensated blocks, and C_{cs} the $N \times 1$ covariance vector between the motion-compensated blocks and the original blocks. (2.13) is a constrained Wiener problem as the scalar weights h_μ sum to 1, i.e., $\mathbf{1}^T h = 1$. A Lagrangian approach leads to the conditional optimal superposition coefficients

$$h = C_{cc}^{-1}\left(C_{cs} - \frac{\mathbf{1}^T C_{cc}^{-1} C_{cs} - 1}{\mathbf{1}^T C_{cc}^{-1}\mathbf{1}}\mathbf{1}\right). \quad (2.14)$$

Experiments with video sequences in [15, 16] reveal that the optimum superposition coefficients are approximately $\frac{1}{N}$ for N superimposed motion-compensated blocks.

Given the entropy code γ and the weighted superposition β, the problem of rate-constrained estimation of superimposed motion is solved by (2.10). The complexity of superimposed motion estimation increases exponentially with the number of superimposed blocks N. An iterative algorithm, which is inspired by the *Iterated Conditional Modes* [120], avoids searching the complete space by successively improving N conditional optimal solutions. Convergence to a local optimum is guaranteed, because the algorithm prohibits an increase of the Lagrangian costs.

0: Assuming N superimposed motion-compensated blocks, the Lagrangian cost function

$$j(c_1, \ldots, c_\mu, \ldots, c_N) = \left\| s - \sum_{\nu=1}^{N} c_\nu h_\nu \right\|_2^2 + \lambda \left| \gamma \left(i[c_1, \ldots, c_N] \right) \right|$$

is subject to minimization for each original block s. Select the entropy code γ, predictor coefficients h, and the Lagrange multiplier λ. Initialize the algorithm with N motion-compensated blocks $(c_1^{(0)}, \ldots, c_N^{(0)})$ and set $k := 0$.

1: Select the μ-th block out of N; start from the first and end with the N-th.

    a: Focus on the μ-th block. All others are kept constant. Select a local neighborhood of the block $c_\mu^{(k)}$ as the conditional search space for block $c_\mu^{(k+1)}$.

    b: Minimize the Lagrangian cost function by full search within the conditional search space and determine the new block $c_\mu^{(k+1)}$.

$$\min_{\{c_\mu^{(k+1)}\}} j(c_1^{(k+1)}, \ldots, c_{\mu-1}^{(k+1)}, c_\mu^{(k+1)}, c_{\mu+1}^{(k)}, \ldots, c_N^{(k)})$$

2: As long as the Lagrangian cost function decreases, continue with step 1 and set $k := k + 1$.

Figure 2.6.   Hypothesis Selection Algorithm for block-based superimposed motion estimation.

The *Hypothesis Selection Algorithm* (HSA) in Fig. 2.6 provides a locally optimal solution for (2.10). The HSA is initialized with $N > 1$ motion-compensated blocks by splitting the optimal motion-compensated block for $N = 1$. The computational complexity of finding a solution for $N = 1$ is rather moderate. This optimal motion-compensated block is repeated N times to generate the initial vector of N motion-compensated blocks. For each of the N motion-compensated blocks in each iteration, the HSA performs a full search within a conditional search space. The size of this conditional search space affects both the quality of the local optimum and the computational complexity of the algorithm [15, 16].

2.3.3 Quantizer Selection at the Residual Encoder

Hybrid video codecs usually encode the motion-compensated prediction error signal. For that, a uniform scalar quantizer with quantizer step-size Q is utilized. Rate-constrained motion estimation raises the question how to select the quantizer step-size dependent on the Lagrange multiplier λ. A solution to this problem is suggested in [121].

Given the Lagrangian cost function $J = D + \lambda R$, total rate R and distortion D are in equilibrium for $dJ = dD + \lambda dR = 0$. Consequently, λ is the negative derivative of the distortion with respect to the total rate.

$$\lambda = -\frac{dD}{dR} \tag{2.15}$$

The rate of the motion-compensated predictor R_p and the rate of the residual encoder R_r sum up to the total rate R. Thus, we can write

$$dD = \frac{\partial D}{\partial R_p} dR_p + \frac{\partial D}{\partial R_r} dR_r \tag{2.16}$$

$$dR = dR_p + dR_r \tag{2.17}$$

and

$$dJ = \left(\frac{\partial D}{\partial R_p} + \lambda\right) dR_p + \left(\frac{\partial D}{\partial R_r} + \lambda\right) dR_r = 0. \tag{2.18}$$

(2.18) is the condition for the equilibrium and has to hold for any dR_p and dR_r. As a result, the partial derivatives of the distortion are identical and equal to $-\lambda$.

$$\lambda = -\frac{dD}{dR} = -\frac{\partial D}{\partial R_p} = -\frac{\partial D}{\partial R_r} \tag{2.19}$$

[122] discusses the optimal rate allocation between motion vector rate and prediction error rate and reports the identity of the partial derivatives. Assuming that the total rate is constant and that an infinitesimal bit has to be assigned, the optimum trade-off is achieved when the decrease in distortion caused by this infinitesimal bit is equal for both motion-compensated prediction and residual encoding.

A memoryless Gaussian signal is assumed for the reconstruction error with the distortion-rate function

$$D(R_p, R_r) = \sigma_e^2(R_p)2^{-2R_r} = \sigma_e^2(R_p)e^{-R_r 2\ln 2}, \qquad (2.20)$$

where $\sigma_e^2(R_p)$ is the variance of the prediction error as a function of the predictor rate R_p. The partial derivative provides $\lambda = D(R_p, R_r)2\ln 2$. Note that the factor $2\ln 2$ is related to the slope of 6 dB/bit. At high rates, the rate of the predictor R_p is negligible compared to the rate of the residual encoder R_r and the distortion $D(R_p, R_r)$ is dominated by the quantizer of the residual encoder. A high-rate approximation for the uniform scalar quantizer

$$D(R_p, R_r) = \frac{Q^2}{12} \qquad (2.21)$$

relates the quantizer step-size Q to the distortion $D(R_p, R_r)$ and λ.

$$\lambda = \frac{\ln 2}{6}Q^2 \approx 0.1Q^2 \qquad (2.22)$$

In practice, this theoretical relationship is modified to

$$\lambda \approx 0.2Q^2. \qquad (2.23)$$

In [121, 123], the functional dependency $\lambda = 0.85 \cdot QP^2$ is suggested for the ITU-T Rec. H.263, where QP is the H.263 quantization parameter. The step-size of the H.263 uniform quantizer is determined by multiplying the quantization parameter by 2, i.e. $Q = 2 \cdot QP$. This result motivates the approximation in (2.23).

2.3.4 Efficient Motion Estimation

Multiframe motion-compensated prediction and, in particular, superimposed motion-compensated prediction increase the computational complexity of motion estimation. Fast search algorithms can be used to reduce the computational burden without sacrificing performance.

[124] presents a fast exhaustive search algorithm for motion estimation. The basic idea is to obtain the best estimate of the motion vectors by successively eliminating the search positions in the search window and thus decreasing the number of matching evaluations that require very intensive computations.

An improved approach is published in [125]. This work proposes a fast block-matching algorithm that uses fast matching error measures besides the conventional mean absolute error or mean square error. An incoming block in the current frame is compared to candidate blocks within the search window using multiple matching criteria. The fast matching error measures are established on the integral projections, having the advantages of being good block features and having simple complexity in measuring matching errors. Most of the candidate blocks can be rejected only by calculating one or more of the fast matching error measures. The time-consuming computations of mean square error or mean absolute error are performed on only a few candidate blocks that first pass all fast matching criteria.

One fast elimination criterion is based on the well-known triangle inequality. A more interesting one is outlined in the following: Let $e_i = s_i - \hat{s}_i$ be the prediction error for pixel $i = 1, 2, \ldots, L$ and L be the number of pixels in a block. s_i are the pixel values of the block in the current frame and $\hat{s}_i$ the pixel values of the block used for prediction. For all error values e_i, the inequality

$$\sum_{i=1}^{L} \left(e_i - \frac{1}{L} \sum_{j=1}^{L} e_j \right)^2 \geq 0 \qquad (2.24)$$

holds and can be simplified to

$$\sum_{i=1}^{L} e_i^2 \geq \frac{1}{L} \left(\sum_{j=1}^{L} e_j \right)^2 . \qquad (2.25)$$

(2.25) states that the sum-square error is always larger or equal to the normalized squared difference of block pixel sums.

$$\sum_{i=1}^{L} (s_i - \hat{s}_i)^2 \geq \frac{1}{L} \left(\sum_{j=1}^{L} s_j - \sum_{j=1}^{L} \hat{s}_j \right)^2 \qquad (2.26)$$

These block pixel sums eliminate efficiently search positions with large errors but do not require time-consuming computations. Moreover, the sum for the incoming block in the current frame $\sum_j s_j$ needs to be calculated only once.

2.4   Theory of Motion-Compensated Prediction

Can the future of a sequence be predicted based on its past? If so, how good could this prediction be? These questions are frequently encountered in many applications that utilize prediction schemes [126]. Video applications that use predictive coding are no exception and several researchers take the journey to explore motion-compensated prediction. For example, CUVELIER and VANDENDORPE investigate the statistical properties of prediction error images [127], analyze coding of interlaced or progressive video [128], and explore motion-compensated interpolation [129]. LI and GONZALES propose a locally quadratic model of the motion estimation error criterion function and apply it for subpixel interpolation [130]. RIBAS-CORBERA and NEUHOFF use their analytical model to optimize motion-vector accuracy in block-based video coding [131]. GULERYUZ and ORCHARD investigate rate-distortion based temporal filtering for video compression [132] and provide a rate-distortion analysis of DPCM compression of Gaussian autoregressive sequences [133]. PANG and TAN examine the role of the optimum loop filter in hybrid coders [134]. WEDI considers aliasing and investigates in [135] a time-recursive interpolation filter for motion compensated prediction. He extends his work to an adaptive interpolation filter [136] and suggests a theoretical basis for adaptive interpolation.

GIROD presents in [137, 12, 138] an approach to characterize motion-compensated prediction. The approach relates the motion-compensated prediction error to the displacement error caused by inaccurate motion compensation. The model utilizes a Gaussian or uniform distributed displacement error to both capture the average accuracy of motion-compensation and evaluate the impact on the motion-compensated prediction error variance. A detailed discussion of fractional-pel motion-compensation and the efficiency of motion-compensated prediction is investigated in [1].

In [139, 11], this theory has been extended to multihypothesis motion-compensated prediction to investigate multiple, linearly combined motion-compensated prediction signals. The paper introduces a statistical model for multiple motion-compensated signals, also called hypotheses. Each hypothesis utilizes the statistical model in [12]. In particular, it assumes statistical independence among the hypotheses. With this approach, the paper continues discussing the optimum Wiener filter for multihypothesis motion-compensated prediction.

The discussion is based on a high-rate approximation of the rate-distortion performance of a hybrid video codec. The high-rate approximation assumes that the residual encoder encodes the prediction error with very small distortion. That is, any reference frame at time instance t that will be used for prediction suffers no degradation and is identical to the original frame at time instance t. With this assumption, the performance of a hybrid video codec can be investigated dependent on the efficiency of motion-compensated prediction.

2.4.1   Frame Signal Model

Let $\mathbf{v} = \{\mathbf{v}[l], l \in \Pi\}$ be a scalar random field over a two-dimensional orthogonal grid Π with horizontal and vertical spacing of 1. The vector $l = (x, y)^T$ denotes a particular location in the lattice Π. We call this a space-discrete frame.

A frame is characterized by its autocorrelation function and power spectral density. The scalar space-discrete cross correlation function [140] is defined according to

$$\phi_{\mathbf{ab}}[l] = E\left\{\mathbf{a}[l_0 + l]\mathbf{b}^*[l_0]\right\} \tag{2.27}$$

where **a** and **b** are complex-valued, jointly wide-sense stationary random fields, **b*** is the complex conjugate of **b**, and $l_0 \in \Pi$ is an arbitrary location. For wide-sense stationary random fields, the correlation function does not depend on l_0 but only on the relative two-dimensional shift l. The cross spectral density is defined according to

$$\Phi_{\mathbf{ab}}(\omega) = \mathcal{F}_*\{\phi_{\mathbf{ab}}[l]\} \tag{2.28}$$

where $\omega = (\omega_x, \omega_y)^T$ is the vector valued frequency and $\mathcal{F}_*\{\cdot\}$ the 2D band-limited discrete-space Fourier transform. In particular, the transform is

$$\Phi_{\mathbf{ab}}(\omega) = \sum_{l\in\Pi} \phi_{\mathbf{ab}}[l]e^{-j\omega^T l} \quad \forall \quad \omega \in \,]-\pi, \pi]\times]-\pi, \pi] \tag{2.29}$$

and its inverse is

$$\phi_{\mathbf{ab}}[l] = \frac{1}{4\pi^2}\int\limits_{-\pi}^{\pi}\int\limits_{-\pi}^{\pi} \Phi_{\mathbf{ab}}(\omega)e^{jl^T\omega}d\omega \quad \forall \quad l \in \Pi. \tag{2.30}$$

It is assumed that an isotropic, exponentially decaying, space-continuous autocorrelation function

$$\phi_{\mathbf{vv}}(x, y) = \sigma_\mathbf{v}^2 \rho_\mathbf{v}^{\sqrt{x^2+y^2}} \tag{2.31}$$

with spatial correlation coefficient $\rho_\mathbf{v}$ and overall signal variance $\sigma_\mathbf{v}^2 = 1$ characterizes a space-continuous frame **v**. If we neglect spectral replications due to sampling, this autocorrelation function corresponds to the isotropic spatial power spectral density

$$\Phi_{\mathbf{vv}}(\omega_x, \omega_y) = \frac{2\pi}{\omega_0^2}\left(1 + \frac{\omega_x^2 + \omega_y^2}{\omega_0^2}\right)^{-\frac{3}{2}} \tag{2.32}$$

with $\omega_0 = -\ln(\rho_\mathbf{v})$. Please refer to Appendix A.4 for details. A spatial correlation coefficient of $\rho_\mathbf{v} = 0.93$ is typical for video signals. Band-limiting the space-continuous signals to the frequencies $]-\pi, \pi]\times]-\pi, \pi]$ and sampling them at the lattice locations Π provides the space-discrete signals.

2.4.2   Signal Model for Motion-Compensated Prediction

For motion-compensated prediction, the motion-compensated signal c is used to predict the current frame s. The model assumes that both the current frame s and the motion-compensated signal c originate from the model frame v. The model frame captures the statistical properties of a frame without motion and residual noise.

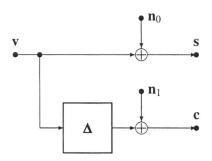

Figure 2.7.   Signal model for the current frame s and the motion-compensated signal c.

Fig. 2.7 depicts the signal model for motion-compensated prediction. Adding statistically independent white Gaussian noise n_0 to the model frame v generates the current frame signal s. Shifting the model frame v by the statistically independent displacement error $\Delta = (\Delta_x, \Delta_y)^T$ and adding statistically independent white Gaussian noise n_1 provides the motion-compensated signal c. For the shift, the ideal reconstruction of the band-limited signal $v[l]$ is shifted by the continuous valued displacement error Δ and re-sampled on the original orthogonal grid. The noise signals n_0 and n_1 are also statistically independent.

The model assumes that the true displacement is known and captures only the displacement error. Obviously, motion-compensated prediction should work best if we compensate the true displacement of the scene exactly for a prediction signal. Less accurate compensation will degrade the performance. To capture the limited accuracy of motion compensation, a vector-valued displacement error Δ is associated with the motion-compensated signal c. The displacement error reflects the inaccuracy of the displacement vector used for motion compensation and transmission. The displacement vector field can never be completely accurate since it has to be transmitted as side information with a limited bit-rate. The model assumes that the 2-D displacement error Δ

is isotropic Gaussian with variance $\sigma_{\mathbf{\Delta}}^2$

$$p_{\mathbf{\Delta}}(\mathbf{\Delta}) = \frac{1}{2\pi\sigma_{\mathbf{\Delta}}^2} e^{-\frac{\mathbf{\Delta}^T\mathbf{\Delta}}{2\sigma_{\mathbf{\Delta}}^2}}. \tag{2.33}$$

With this assumption, the accuracy of motion compensation is now captured by the displacement error variance $\sigma_{\mathbf{\Delta}}^2$. Further, it is assumed that the displacement error is entirely due to rounding and is uniformly distributed in the interval $[-2^{\beta-1}, 2^{\beta-1}] \times [-2^{\beta-1}, 2^{\beta-1}]$, where $\beta = 0$ for integer-pel accuracy, $\beta = -1$ for half-pel accuracy, $\beta = -2$ for quarter-pel accuracy, etc. Given the displacement inaccuracy β, the displacement error variance is

$$\sigma_{\mathbf{\Delta}}^2 = \frac{2^{2\beta}}{12}. \tag{2.34}$$

The current frame **s** is linearly predicted from the motion-compensated signal **c**. The prediction error is defined by

$$\mathbf{e}[l] = \mathbf{s}[l] - f[l] * \mathbf{c}[l] \tag{2.35}$$

where $f[l]$ is the impulse response of the 2-D prediction filter. The asterisk $*$ denotes 2-D convolution on the original orthogonal grid Π. The filter is determined according to the minimum mean square error criterion. The normalized power spectral density of the minimum prediction error is determined by

$$\frac{\Phi_{\mathbf{ee}}(\omega)}{\Phi_{\mathbf{ss}}(\omega)} = 1 - \frac{1}{1+\alpha_0(\omega)} \cdot \frac{P(\omega)P^*(\omega)}{1+\alpha_1(\omega)} \tag{2.36}$$

where $P(\omega)$ is the 2-D continuous-space Fourier transform $\mathcal{F}\{p_{\mathbf{\Delta}}(\mathbf{\Delta})\}$ of the 2-D displacement error PDF $p_{\mathbf{\Delta}}(\mathbf{\Delta})$. $\alpha_0(\omega)$ and $\alpha_1(\omega)$ are the normalized power spectral densities of the residual noise in the current frame **s** and in the motion-compensated signal **c**, respectively.

$$\alpha_\mu(\omega) = \frac{\Phi_{\mathbf{n}_\mu\mathbf{n}_\mu}(\omega)}{\Phi_{\mathbf{vv}}(\omega)} \tag{2.37}$$

Please note, that the power spectral density of the minimum prediction error is normalized to that of the current frame $\Phi_{\mathbf{ss}}(\omega)$, whereas the power spectral density of the residual noise is normalized to that of the model frame $\Phi_{\mathbf{vv}}(\omega)$.

2.4.3 Signal Model for Multihypothesis Prediction

For multihypothesis motion-compensated prediction, N motion-compensated signals $\mathbf{c}_\mu$ with $\mu = 1, 2, \ldots, N$ are used to predict the current frame **s**.

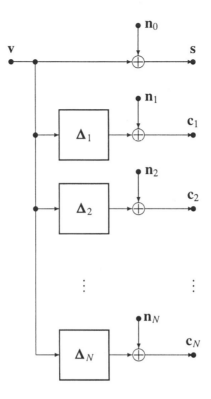

Figure 2.8. Signal model for the current frame **s** and N motion-compensated prediction signals (hypotheses) $\mathbf{c}_\nu$.

The model assumes that both the current frame **s** and N motion-compensated signals $\mathbf{c}_\mu$ originate from the model frame **v**. The model frame captures the statistical properties of a frame without motion and residual noise.

 Fig. 2.8 depicts the signal model for multihypothesis motion-compensated prediction. Adding statistically independent white Gaussian noise $\mathbf{n}_0$ to the model frame **v** generates the current frame signal **s**. Shifting the model frame **v** by the statistically independent displacement error $\boldsymbol{\Delta}_\mu$ and adding statistically independent white Gaussian noise $\mathbf{n}_\mu$ provides the μ-th motion-compensated signal $\mathbf{c}_\mu$. For the shift, the ideal reconstruction of the band-limited signal $v[l]$ is shifted by the continuous valued displacement error Δ_μ and re-sampled on the original orthogonal grid. The noise signals $\mathbf{n}_\mu$ and $\mathbf{n}_\nu$ with $\mu, \nu = 0, 1, \ldots, N$ are mutually statistically independent for $\mu \neq \nu$.

 The model assumes that N true displacements exist and utilizes N displacement errors to capture the limited accuracy of the N motion-compensated signals. For that, a vector-valued displacement error $\boldsymbol{\Delta}_\mu$ with $\mu = 1, 2, \ldots, N$

is associated with the μ-th motion-compensated signal $\mathbf{c}_\mu$. The μ-th displacement error reflects the inaccuracy of the μ-th displacement vector used for multihypothesis motion compensation. The model assumes that all displacement errors $\boldsymbol{\Delta}_\mu$ with $\mu = 1, 2, \ldots, N$ are isotropic Gaussian with identical variance σ_Δ^2 according to (2.33). The displacement errors $\boldsymbol{\Delta}_\mu$ and $\boldsymbol{\Delta}_\nu$ with $\mu, \nu = 1, 2, \ldots, N$ are mutually statistically independent for $\mu \neq \nu$.

The current frame $\mathbf{s}$ is linearly predicted from the vector of motion-compensated signals $\mathbf{c} = (\mathbf{c}_1, \mathbf{c}_2, \ldots, \mathbf{c}_N)^T$. The scalar prediction error is defined by

$$\mathbf{e}[l] = \mathbf{s}[l] - f[l] * \mathbf{c}[l] \tag{2.38}$$

with the row-vector of impulse responses $f[l] = (f_1[l], f_2[l], \ldots, f_N[l])$ of the 2-D prediction filter. The asterisk $*$ denotes 2-D convolution on the original orthogonal grid Π according to

$$f[l] * \mathbf{c}[l] = \sum_{l_0 \in \Pi} f[l_0] \mathbf{c}[l - l_0]. \tag{2.39}$$

The filter is determined according to the minimum mean square error criterion. The normalized power spectral density of the minimum prediction error is determined by

$$\frac{\Phi_{ee}}{\Phi_{ss}} = \tag{2.40}$$

$$1 - \frac{1}{1 + \alpha_0} \begin{pmatrix} P_1 \\ P_2 \\ \vdots \\ P_N \end{pmatrix}^H \begin{pmatrix} 1 + \alpha_1 & P_1 P_2^* & \cdots & P_1 P_N^* \\ P_2 P_1^* & 1 + \alpha_2 & \cdots & P_2 P_N^* \\ \vdots & \vdots & \ddots & \vdots \\ P_N P_1^* & P_N P_2^* & \cdots & 1 + \alpha_N \end{pmatrix}^{-1} \begin{pmatrix} P_1 \\ P_2 \\ \vdots \\ P_N \end{pmatrix}.$$

For simplicity, the argument ω is omitted. $P_\mu(\omega)$ is the 2-D continuous-space Fourier transform $\mathcal{F}\{p_{\boldsymbol{\Delta}_\mu}(\Delta)\}$ of the μ-th 2-D displacement error PDF $p_{\boldsymbol{\Delta}_\mu}(\Delta)$. $\alpha_0(\omega)$ and $\alpha_\mu(\omega)$ with $\mu = 1, 2, \ldots, N$ are the normalized power spectral densities of the residual noise in the current frame $\mathbf{s}$ and in the motion-compensated signals $\mathbf{c}_\mu$, respectively. The normalized power spectral density of the residual noise is defined according to (2.37). Please note, that the power spectral density of the minimum prediction error is normalized to that of the current frame $\Phi_{ss}(\omega)$, whereas the power spectral density of the residual noise is normalized to that of the model frame $\Phi_{vv}(\omega)$.

2.4.4   Performance Measures

With high-rate assumptions, the motion-compensated prediction error $\mathbf{e}$ is sufficient for performance evaluation. As the spatial correlation of the predic-

tion error is only weak, the potential for redundancy reduction in the residual
encoder is relatively small [11]. This suggests that the prediction error variance

$$\sigma_e^2 = \frac{1}{4\pi^2} \int\limits_{-\pi}^{\pi} \int\limits_{-\pi}^{\pi} \Phi_{ee}(\omega)d\omega \tag{2.41}$$

is a useful measure that is related to the minimum achievable transmission bit-
rate for a given signal-to-noise ratio [141]. The minimization of the prediction
error variance (2.41) is widely used to obtain the displacement vector and control the coding mode in practical systems. A more refined measure is the rate
difference [142]

$$\Delta R = \frac{1}{8\pi^2} \int\limits_{-\pi}^{\pi} \int\limits_{-\pi}^{\pi} \log_2 \left(\frac{\Phi_{ee}(\omega)}{\Phi_{ss}(\omega)} \right) d\omega. \tag{2.42}$$

In (2.42), $\Phi_{ee}(\omega)$ and $\Phi_{ss}(\omega)$ are the power spectral densities of the prediction error e and the current frame s, respectively. Unlike (2.41), the rate difference (2.42) takes the spatial correlation of the prediction error e and the
original signal s into account. It represents the maximum bit-rate reduction (in
bits/sample) possible by optimum encoding of the prediction error e, compared
to optimum intra-frame encoding of the signal s, for Gaussian wide-sense stationary signals for the same mean square reconstruction error [141]. A negative
ΔR corresponds to a reduced bit-rate compared to optimum intra-frame coding, while a positive ΔR is a bit-rate increase due to motion-compensation, as
it can occur for inaccurate motion-compensation. The maximum bit-rate reduction can be fully realized at high bit-rates, while for low bit-rates the actual
gain is smaller. The rate required for transmitting the displacement estimate is
neglected. The optimum balance between rates for the prediction error signal
and displacement vectors strongly depends on the total bit-rate. For high rates,
it is justified to neglect the rate for the displacement vectors [11].

2.4.5   Conclusions

Based on the simplifying assumptions, several important conclusions can be
drawn from the theory. Doubling the accuracy of motion compensation, such
as going from integer-pel to 1/2-pel accuracy, can reduce the bit-rate by up to
1 bit per sample independent of N for the noise-free case. An optimum combination of N hypotheses always lowers the bit-rate for increasing N. If each
hypotheses is equally good in terms of displacement error PDF, doubling N
can yield a gain of 0.5 bits per sample if there is no residual noise. If realistic
residual noise levels are taken into account, the gains possible by doubling the

number of hypotheses, N, decreases with increasing N. Diminishing returns and, ultimately, saturation is observed. If the residual noise power increases, doubling and ultimately quadrupling the number of hypotheses N becomes more efficient than doubling the accuracy of motion compensation. The critical accuracy beyond which the gain due to more accurate motion compensation is small moves to larger displacement error variances with increasing noise and increasing number of hypotheses N. Hence, sub-pel accurate motion compensation becomes less important with multihypothesis motion-compensated prediction. Spatial filtering of the motion-compensated candidate signals becomes less important if more hypotheses are combined [11].

2.5 Three-Dimensional Subband Coding of Video

Hybrid video coding schemes utilize predictive coding with motion-compensated prediction for efficient compression. Such compression schemes require sequential processing of video signals which makes it difficult to achieve efficient embedded representations. A multiresolution signal decomposition of the video signal seems to be promising to achieve efficient embedded representations. MALLAT [148] discusses the wavelet representation as a suitable tool for multiresolution signal decomposition. Moreover, wavelets are also suitable for coding applications [149, 150]. For example, SHAPIRO proposes zerotrees of wavelet coefficients for embedded image coding [151]. USEVITCH derives optimal bit allocations for biorthogonal wavelet coders which result in minimum reconstruction error [152]. TAUBMAN proposes a new image compression algorithm based on independent *Embedded Block Coding with Optimized Truncation* of the embedded bit-stream (EBCOT) [153].

Practical wavelet coding schemes are characterized by the construction of the wavelets. DAUBECHIES and SWELDENS factor wavelet transforms into lifting steps [154–156]. This construction scheme permits wavelet transforms that map integers to integers [157], a desirable property for any practical coding scheme. Moreover, the lifting scheme permits also adaptive wavelet transforms [158, 159]. Adaptive schemes like motion compensation can be incorporated into the lifting scheme.

2.5.1 Motion Compensation and Subband Coding

For hybrid video coding, there are attempts to keep the efficient predictive architecture with motion compensation and to use wavelet-based techniques for coding the displaced frame difference, e.g. [160]. These are spatial subband coding techniques and do not provide three-dimensional subbands. Early attempts of three-dimensional subband coding did not employ motion compensation. KNAUER applied in [18] the Hadamard transform for real-time television compression by considering the sequence of frames as a volume. Further research is published by VETTERLI et al. [161], BIEMOND et al. [162], and JAYANT et al. [163]. Interestingly, ZHANG et al. investigate memory-constrained 3D wavelet transforms without boundary effects and use the lifting structure [164, 165]. BARLAUD et al. suggest a 3D scan based wavelet transform with reduced memory requirements [166]. 3D wavelet coding with arbitrary regions of support is discussed by MINAMI et al. [167].

Three-dimensional coding with motion compensation has first been suggested by KRONANDER [19, 168]. AKIYAMA et al. use global motion compensation and three-dimensional transform coding [169]. Global motion compensation with 3D wavelet coding is also investigated by CHOU et al. in [170]. ZHANG and ZAFAR present a motion-compensated wavelet transform coder for color video compression [171]. OHM starts his investigation of 3D subband video coding [24, 172] with integer-pel accurate motion compensation and first order quadrature mirror filter in the temporal domain [173, 174]. WOODS et al. discuss a scheme for object-based spatio-temporal subband coding [175]. They optimize the trade-off of the rate between motion vectors and 3-D subbands [25] and consider digital cinema applications [176]. A bit allocation scheme for subband compression of HDTV is published in [177]. For video communication over wireless channels, CHOU and CHEN propose a perceptually optimized 3-D subband codec [178].

The multiresolution signal decomposition of the video signal permits temporal, spatial, and rate-distortion scalability. This important feature is investigated by many researchers. UZ et al. suggest a scheme for interpolative multiresolution coding of digital HDTV [179, 180]. In addition to the multiresolution representation of the motion-compensated three-dimensional signal, researchers investigate also the multiresolution representation of motion, like OHM [181] and ZHANG et al. [182]. TAUBMAN and ZAKHOR propose a common framework for rate and distortion based scaling of highly scalable compressed video [183, 184, 23, 185, 186]. PEARLMAN et al. introduce in [187, 188] an embedded wavelet video codec using three-dimensional set partitioning in hierarchical trees [189]. WOODS et al. discuss rate-constrained multiresolution transmission of video [190], and present a resolution and frame-rate scalable subband video coder [191]. RANGANATH et al. outline a highly scalable wavelet-based codec for very low bit-rate environments and introduces tri-zerotrees [192]. PESQUET-POPESCU et al. suggest a method for context modeling in the spatio-temporal trees of wavelet coefficients [193] and propose the strategy of fully scalable zerotree coding [194]. ZHANG et al. investigate three-dimensional embedded block coding with optimized truncation [195].

2.5.2 Motion-Compensated Lifted Wavelets

The previously mentioned lifting scheme permits adaptive wavelet transforms. For the temporal subband decomposition, PESQUET-POPESCU and BOTTREAU improve the compression efficiency with lifting schemes that use motion compensation [196]. ZHANG et al. also incorporate motion compensation into the lifting scheme [197] and use this method for 3D wavelet compression of concentric mosaics [198]. SECKER and TAUBMAN investigate lifting schemes with block motion compensation [20] and deformable mesh motion compensation [199]. OHM reviews the recent progress and discusses novel aspects with respect to irregular sampled signals, shift variance of wavelet transforms and non-dyadic wavelet processing [200]. PATEUX et al. investigate several motion-compensated lifting implementations and compare to standardized hybrid codecs [201]. BARLAUD et al. extend their 3D scan-based wavelet transform codec and use the motion-compensated lifting scheme [202].

Chapter 3

MOTION-COMPENSATED PREDICTION WITH COMPLEMENTARY HYPOTHESES

3.1 Introduction

As discussed in Section 2.4, the theoretical investigation in [11] shows that a linear combination of multiple motion-compensated signals can improve the performance of motion-compensated prediction for video coding. This chapter extends that work by introducing the concept of complementary hypotheses [203, 204].

To motivate this concept, let us consider pairs of motion-compensated signals. The two signals are simply averaged to form the prediction signal. We ask the question what kind of pairs are necessary to achieve the best prediction performance of superimposed motion compensation. If a pair consists of two identical hypotheses, the superimposed prediction signal is identical to either one of the hypotheses and we expect no improvement over motion-compensated prediction with just one hypothesis. But, in general, there will be pairs of hypotheses that outperform motion-compensated prediction with single hypotheses. Our approach is to model the dependency among the two signals by a correlation coefficient and investigate its impact on the performance of superimposed motion compensation.

Our assumption that there will be N-tuples of hypotheses that outperform motion-compensated prediction with single hypotheses is supported by experimental results. As discussed in Section 2.3.2, the work in [15, 16] demonstrates experimentally that such efficient N-tuples exist. The work proposes an iterative algorithm for block-based rate-constrained superimposed motion estimation. The algorithm improves conditionally optimal solutions and provides a local optimum for the joint estimation problem. The results demonstrate that joint estimation of hypotheses is important for superimposed motion-compensated prediction.

The outline of this chapter is as follows: Section 3.2 extends the known model of multihypothesis motion-compensated prediction. Correlated displacement error for superimposed prediction are discussed and the concept of motion compensation with complementary hypotheses is introduced. Further, we discuss the gradient of the prediction error variance for superimposed motion-compensated signals and the impact of a particular frame signal model. Section 3.3 analyzes "noisy" hypotheses, investigates both averaging and Wiener filtering, and provides performance results for jointly estimated hypotheses. Section 3.4 explores hypothesis switching as a method to select efficient hypotheses for superimposed motion-compensated prediction. First, the signal model for forward-adaptive hypothesis switching is introduced. Second, the problem is approached by minimizing the radial displacement error. Third, a property of the assumed PDF allows the definition of an equivalent predictor. And finally, forward-adaptive hypothesis switching is combined with superimposed motion-compensated prediction. Section 3.5 discusses image sequence coding where individual images are predicted with a varying number of hypotheses. The impact of superimposed motion estimation on the overall coding efficiency is investigated.

3.2 Extended Model for Superimposed Motion-Compensated Prediction

3.2.1 Superimposed Prediction and Correlated Displacement Error

We extend the model for multihypothesis motion-compensated prediction as discussed in Section 2.4 such that correlated displacement errors can be investigated. Let $\mathbf{s}[l]$ and $\mathbf{c}_\mu[l]$ be scalar two-dimensional signals sampled on an orthogonal grid with horizontal and vertical spacing of 1. The vector $l = (x, y)^T$ denotes the location of the sample. For the problem of superimposed motion compensation, we interpret $\mathbf{c}_\mu$ as the μ-th of N motion-compensated signals available for prediction, and $\mathbf{s}$ as the current frame to be predicted. We also call $\mathbf{c}_\mu$ the μ-th hypothesis.

Motion-compensated prediction should work best if we compensate the true displacement of the scene exactly for a prediction signal. Less accurate compensation will degrade the performance. To capture the limited accuracy of motion compensation, we associate a vector-valued displacement error $\mathbf{\Delta}_\mu$ with the μ-th hypothesis $\mathbf{c}_\mu$. The displacement error reflects the inaccuracy of the displacement vector used for motion compensation and transmission.

The displacement vector field can never be completely accurate since it has to be transmitted as side information with a limited bit-rate. For simplicity, we assume that all hypotheses are shifted versions of the current frame signal **s**. The shift is determined by the vector-valued displacement error Δ_μ of the μ-th hypotheses. For that, the ideal reconstruction of the band-limited signal $s[l]$ is shifted by the continuous valued displacement error and re-sampled on the original orthogonal grid. For now, the translatory displacement model as depicted in Fig. 2.8 omits "noisy" signal components.

A superimposed motion-compensated predictor forms a prediction signal by averaging N hypotheses $c_\mu[l]$ in order to predict the current frame signal $s[l]$. The prediction error for each pel at location l is the difference between the current frame signal and N averaged hypotheses

$$e[l] = s[l] - \frac{1}{N} \sum_{\mu=1}^{N} c_\mu[l]. \tag{3.1}$$

Assume that **s** and c_μ are generated by a jointly wide-sense stationary random process with the real-valued scalar two-dimensional power spectral density $\Phi_{ss}(\omega)$ as well as the cross spectral densities $\Phi_{c_\mu s}(\omega)$ and $\Phi_{c_\mu c_\nu}(\omega)$. Power spectra and cross spectra are defined according to (2.28) where $\omega = (\omega_x, \omega_y)^T$ is the vector valued frequency.

The power spectral density of the prediction error in (3.1) is given by the power spectrum of the current frame and the cross spectra of the hypotheses

$$\Phi_{ee}(\omega) = \Phi_{ss}(\omega) - \frac{2}{N} \sum_{\mu=1}^{N} \Re\left\{ \Phi_{c_\mu s}(\omega) \right\} + \frac{1}{N^2} \sum_{\mu=1}^{N} \sum_{\nu=1}^{N} \Phi_{c_\mu c_\nu}(\omega), \tag{3.2}$$

where $\Re\{\cdot\}$ denotes the real component of the, in general, complex valued cross spectral densities $\Phi_{c_\mu s}(\omega)$. We adopt the expressions for the cross spectra from [11], where the displacement errors Δ_μ are interpreted as random variables which are statistically independent from **s**:

$$\Phi_{c_\mu s}(\omega) = \Phi_{ss}(\omega) E\left\{ e^{-j\omega^T \Delta_\mu} \right\} \tag{3.3}$$

$$\Phi_{c_\mu c_\nu}(\omega) = \Phi_{ss}(\omega) E\left\{ e^{-j\omega^T (\Delta_\mu - \Delta_\nu)} \right\} \tag{3.4}$$

Like in [11], we assume a power spectrum Φ_{ss} that corresponds to an exponentially decaying isotropic autocorrelation function with a correlation coefficient ρ_s.

For the μ-th displacement error Δ_μ, a 2-D stationary normal distribution with variance σ_Δ^2 and zero mean is assumed where the x- and y-components are statistically independent. The displacement error variance is the same for all N hypotheses. This is reasonable because all hypotheses are compensated with the same accuracy. Further, the pairs (Δ_μ, Δ_ν) are assumed to be jointly Gaussian random variables. The predictor design in [16] showed that there is no preference among the N hypotheses. Consequently, the correlation coefficient ρ_Δ between two displacement error components $\Delta_{x\mu}$ and $\Delta_{x\nu}$ is the same for all pairs of hypotheses. The above assumptions are summarized by the covariance matrix of a displacement error component.

$$C_{\Delta_x \Delta_x} = \sigma_\Delta^2 \begin{pmatrix} 1 & \rho_\Delta & \cdots & \rho_\Delta \\ \rho_\Delta & 1 & \cdots & \rho_\Delta \\ \vdots & \vdots & \ddots & \vdots \\ \rho_\Delta & \rho_\Delta & \cdots & 1 \end{pmatrix}. \tag{3.5}$$

Since the covariance matrix is nonnegative definite [205], the correlation coefficient ρ_Δ in (3.5) has the limited range

$$\frac{1}{1-N} \leq \rho_\Delta \leq 1 \quad \text{for} \quad N = 2, 3, 4, \ldots, \tag{3.6}$$

which is dependent on the number of hypotheses N. More details on this result can be found in Appendix A.1. In contrast to [11], we do not assume that the displacement errors Δ_μ and Δ_ν are mutually independent for $\mu \neq \nu$.

These assumptions allow us to express the expected values in (3.3) and (3.4) in terms of the 2-D Fourier transform P of the continuous 2-D probability density function of the displacement error Δ_μ.

$$\begin{aligned} E\left\{e^{-j\omega^T \Delta_\mu}\right\} &= \int_{\mathcal{R}^2} p_{\Delta_\mu}(\Delta) e^{-j\omega^T \Delta} d\Delta \\ &= P(\omega, \sigma_\Delta^2) \\ &= e^{-\frac{1}{2}\omega^T \omega \sigma_\Delta^2} \end{aligned} \tag{3.7}$$

The expected value in (3.4) contains differences of jointly Gaussian random variables. The difference of two jointly Gaussian random variables is also Gaussian. As the two random variables have equal variance σ_Δ^2, the variance of the difference signal is given by $\sigma^2 = 2\sigma_\Delta^2(1 - \rho_\Delta)$. Therefore, we obtain for the expected value in (3.4)

$$E\left\{e^{-j\omega^T(\Delta_\mu - \Delta_\nu)}\right\} = P\left(\omega, 2\sigma_\Delta^2(1 - \rho_\Delta)\right) \quad \text{for} \quad \mu \neq \nu. \tag{3.8}$$

For $\mu = \nu$, the expected value in (3.4) is equal to one. With that, we obtain for the power spectrum of the prediction error in (3.2):

$$\frac{\Phi_{ee}(\omega)}{\Phi_{ss}(\omega)} = \frac{N+1}{N} - 2P(\omega, \sigma_\Delta^2) + \frac{N-1}{N} P\left(\omega, 2\sigma_\Delta^2(1 - \rho_\Delta)\right) \tag{3.9}$$

Setting $\rho_\Delta = 0$ provides a result which is presented in [11], equation (23), with negligible noise $\alpha_\mu = 0$, averaging filter F, and identical characteristic functions $P_\mu = P$.

3.2.2   Complementary Hypotheses

The previous section shows that the displacement error correlation coefficient influences the performance of superimposed motion compensation. An ideal superimposed motion estimator will select sets of hypotheses that optimize the performance of superimposed motion compensation. In the following, we focus on the relationship between the prediction error variance

$$\sigma_e^2 = \frac{1}{4\pi^2} \int\limits_{-\pi}^{\pi} \int\limits_{-\pi}^{\pi} \Phi_{ee}(\omega) d\omega \tag{3.10}$$

and the displacement error correlation coefficient. The prediction error variance is a useful measure because it is related to the minimum achievable transmission bit-rate [11].

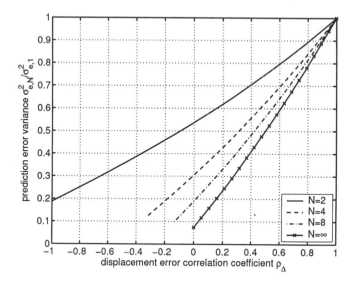

Figure 3.1.   Normalized prediction error variance for superimposed MCP over the displacement error correlation coefficient ρ_Δ. Reference is the single hypothesis predictor. The hypotheses are averaged and no residual noise is assumed. The variance of the displacement error is set to $\sigma_\Delta^2 = 1/12$.

Fig. 3.1 depicts the dependency of the normalized prediction error variance on the displacement error correlation coefficient ρ_Δ within the range (3.6). The dependency is plotted for $N = 2, 4, 8,$ and ∞ for integer-pel accurate motion compensation ($\sigma_\Delta^2 = 1/12$). The correlation coefficient of the frame signal $\rho_s = 0.93$ [11]. Reference is the prediction error variance of the single hypothesis predictor $\sigma_{e,1}^2$. We observe that a decreasing correlation coefficient lowers the prediction error variance. (3.9) implies that this observation holds for any displacement error variance. Fig. 3.1 shows also that identical displacement errors ($\rho_\Delta = 1$) do not reduce the prediction error variance compared to single hypothesis motion compensation. This is reasonable when we consider identical hypotheses. They do not improve superimposed motion-compensation because they have identical displacement errors.

To determine the performance bound, we assume the existence of all N-tuple of hypotheses that obey (3.6) and that an *ideal superimposed motion estimator* is able to determine any desired N-tuple. Assuming a mean square error measure, the optimal ideal superimposed motion estimator minimizes the summed squared error

$$\min \frac{1}{|\mathcal{L}|} \sum_{l \in \mathcal{L}} e^2[l] \tag{3.11}$$

and the expected value

$$\min E \left\{ \frac{1}{|\mathcal{L}|} \sum_{l \in \mathcal{L}} \mathbf{e}^2[l] \right\}, \tag{3.12}$$

where $\mathbf{e}[l]$ denotes the prediction error at pixel location l. In addition, we assume a stationary error signal such that

$$E \left\{ \mathbf{e}^2[l] \right\} = \sigma_e^2[l] = \sigma_e^2 \quad \forall l. \tag{3.13}$$

Consequently, this optimal ideal estimator minimizes the prediction error variance.

$$\min \sigma_e^2 \tag{3.14}$$

Further, σ_e^2 increases monotonically for increasing ρ_Δ. This is a property of (3.9) which is also depicted in Fig. 3.1. The minimum of the prediction error variance is achieved at the lower bound of ρ_Δ.

$$\min_{\frac{1}{1-N} \le \rho_\Delta \le 1} \rho_\Delta \tag{3.15}$$

That is, an optimal ideal superimposed motion estimator minimizes the prediction error variance by minimizing the displacement error correlation coefficient. Its minimum is given by the lower bound of the range (3.6).

$$\rho_\Delta = \frac{1}{1 - N} \quad \text{for} \quad N = 2, 3, 4, \dots \tag{3.16}$$

This insight implies an interesting result for the case $N = 2$: Pairs of hypotheses that operate on the performance bound show the property that their displacement errors are maximally negatively correlated. Hence, the combination of two complementary hypotheses is more efficient than two independent hypotheses.

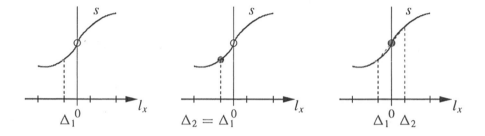

Figure 3.2. Interpolation and displacement error. Due to an inaccurate displacement, only the signal value at spatial location Δ_1 is available (left). Averaging two hypotheses with identical displacement errors does not improve the approximation (middle). When we pick the signal value at spatial location $\Delta_2 = -\Delta_1$ and average the two signal values, we will get closer to the signal value at spatial location $l_x = 0$ (right).

Let us consider the one-dimensional example in Fig. 3.2 where the intensity signal s is a continuous function of the spatial location l_x. A signal value that we want to use for prediction is given at spatial location $l_x = 0$. Due to an inaccurate displacement, only the signal value at spatial location $l_x = \Delta_1$ is available. We assume that the intensity signal is smooth around $l_x = 0$ and not spatially constant. When we pick the signal value at spatial location $l_x = \Delta_2 = -\Delta_1$ and average the two signal values, we will get closer to the signal value at spatial location $l_x = 0$. If we consider many displacement error values Δ_1 with distribution p_Δ, we get for the random variables $\Delta_1 = -\Delta_2$. This results in $\rho_\Delta = -1$.

Fig. 3.3 depicts the rate difference ΔR for multihypothesis motion-compensated prediction over the displacement inaccuracy β for statistically independent displacement errors according to [11]. The rate difference according to (2.42) represents the maximum bit-rate reduction (in bits/sample) possible by optimal encoding of the prediction error $\mathbf{e}$, compared to optimum intra-frame encoding of the signal $\mathbf{s}$ for Gaussian wide-sense stationary signals for the same mean square reconstruction error. A negative rate difference ΔR corresponds to a reduced bit-rate compared to optimum intra-frame coding. The horizontal axis in Fig. 3.3 is calibrated by $\beta = \log_2(\sqrt{12}\sigma_\Delta)$, where $\beta = 0$ for integer-pel accuracy, $\beta = -1$ for half-pel accuracy, $\beta = -2$ for

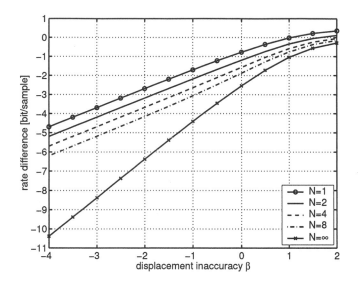

Figure 3.3. Rate difference for superimposed MCP over the displacement inaccuracy β for statistically independent displacement errors. The hypotheses are averaged and no residual noise is assumed.

quarter-pel accuracy, etc [11]. The displacement error variance is given by (2.34).

We observe in Fig. 3.3 that for the case $N = 1$ the slope reaches 1 bit per sample per inaccuracy step. This can also be observed in (3.9) for $N = 1$ when we apply a Taylor series expansion of first order for the function P.

$$\frac{\Phi_{ee}(\omega)}{\Phi_{ss}(\omega)} \approx \sigma_\Delta^2 \omega^T \omega \quad \text{for} \quad \sigma_\Delta^2 \to 0, N = 1 \qquad (3.17)$$

Inserting this result in (2.42) supports the observation in Fig. 3.3

$$\Delta R \approx \beta + const. \quad \text{for} \quad \sigma_\Delta^2 \to 0, N = 1. \qquad (3.18)$$

We observe also in Fig. 3.3 that doubling the number of hypotheses decreases the bit-rate up to 0.5 bits per sample and the slope reaches up to 1 bit per sample per inaccuracy step. The case $N \to \infty$ achieves a slope up to 2 bits per sample per inaccuracy step. This can also be observed in (3.9) for $N \to \infty$ when we apply a Taylor series expansion of second order for the function P

$$\frac{\Phi_{ee}(\omega)}{\Phi_{ss}(\omega)} \approx \frac{1}{4}\sigma_\Delta^4 \left(\omega^T \omega\right)^2 \quad \text{for} \quad \sigma_\Delta^2 \to 0, N \to \infty, \rho_\Delta = 0. \qquad (3.19)$$

Inserting this result in (2.42) supports the observation in Fig. 3.3

$$\Delta R \approx 2\beta + const. \quad \text{for} \quad \sigma_\Delta^2 \to 0, N \to \infty, \rho_\Delta = 0. \qquad (3.20)$$

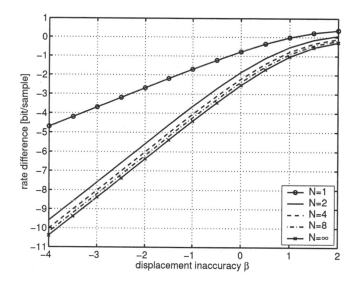

Figure 3.4. Rate difference for superimposed MCP over the displacement inaccuracy β for optimized displacement error correlation. The hypotheses are averaged and no residual noise is assumed.

Fig. 3.4 depicts the rate difference for superimposed motion-compensated prediction over the displacement inaccuracy β for optimized displacement error correlation according to (3.16). We observe for accurate motion compensation that the slope of the rate difference of 2 bits per sample per inaccuracy step is already reached for $N = 2$. For increasing number of hypotheses the rate difference converges to the case $N \to \infty$ at constant slope. This can also be observed in (3.9) when the displacement error correlation coefficient is set to $\rho_\Delta = \frac{1}{1-N}$, and a Taylor series expansion of second order for the function $P(\omega)$ is applied.

$$\frac{\Phi_{ee}(\omega)}{\Phi_{ss}(\omega)} \approx \frac{1}{4}\sigma_\Delta^4(\omega^T\omega)^2\frac{N+1}{N-1} \quad \text{for} \quad \sigma_\Delta^2 \to 0, N = 2, 3, \ldots \quad (3.21)$$

Inserting this result in 2.42 supports the observation in Fig. 3.4 that for $N = 2, 3, \ldots$ the slope reaches up to 2 bits per sample per inaccuracy step.

$$\Delta R \approx 2\beta + \frac{1}{2}\log_2\left(\frac{N+1}{N-1}\right) + const. \quad \text{for} \quad \sigma_\Delta^2 \to 0, N = 2, 3, \ldots \quad (3.22)$$

For very accurate motion compensation $\sigma_\Delta^2 \to 0$ and $N = 2, 3, \ldots$, doubling the number of hypotheses results in a rate difference of

$$\Delta R_2 := \Delta R(2N) - \Delta R(N) \approx \frac{1}{2}\log_2\left(\frac{2N^2 - N - 1}{2N^2 + N - 1}\right). \quad (3.23)$$

For a very large number of hypotheses, the rate difference for doubling the number of hypotheses ΔR_2 converges to zero. Consequently, the prediction gain by optimum noiseless multihypotheses prediction with averaging filter is limited and the rate difference converges to the case $N \to \infty$ at constant slope.

We obtain for the band-limited frame signal the following result: the gain of superimposed motion-compensated prediction with jointly optimal motion estimation over motion-compensated prediction increases by improving the accuracy of motion compensation for each hypothesis. The theoretical results suggest that a practical video coding algorithm should utilize two jointly estimated hypotheses. Experimental results also suggest that the gain by superimposed prediction is limited and that two jointly estimated hypotheses provide a major portion of this achievable gain.

3.2.3   Gradient of the Prediction Error Variance

We observe in Fig. 3.1 that the prediction error variance σ_e^2 increases monotonically for increasing displacement error correlation coefficient ρ_Δ. In the following, we investigate this in more detail and show that this dependency is independent of a particular frame signal model, i.e., a particular frame autocorrelation function.

Again, let $\mathbf{s}(l)$ and $\mathbf{c}_\mu(l)$ be generated by a jointly wide-sense stationary, two-dimensional random process. The vector $l = (x, y)^T$ denotes the location in $\mathcal{R}^2$. Let $\mathbf{c}_\mu$ be the μ-th of N motion-compensated signals available for prediction, and $\mathbf{s}$ be the current frame to be predicted. The limited accuracy of motion compensation is captured by associating a vector-valued displacement error Δ_μ with the μ-th hypothesis $\mathbf{c}_\mu$. For simplicity, we assume that all hypotheses are shifted versions of the current frame signal $\mathbf{s}$. The shift is determined by the vector-valued displacement error Δ_μ of the μ-th hypotheses such that $\mathbf{c}_\mu(l) = \mathbf{s}(l - \Delta_\mu)$.

The superimposed motion-compensated predictor forms a prediction signal by averaging N hypotheses $\mathbf{c}_\mu(l)$ in order to predict the current frame signal $\mathbf{s}(l)$. The prediction error at location l is the difference between the current frame signal and N averaged hypotheses

$$\mathbf{e}(l) = \mathbf{s}(l) - \frac{1}{N} \sum_{\mu=1}^{N} \mathbf{s}(l - \Delta_\mu). \tag{3.24}$$

As we assume wide-sense stationary signals, the prediction error variance $\sigma_e^2 = E\left\{e^2(l)\right\}$ is independent of the location l

$$\sigma_e^2 = \sigma_s^2 - \frac{2}{N}\sum_{\mu=1}^{N} E\left\{\phi_{ss}(\Delta_\mu)\right\} + \frac{1}{N^2}\sum_{\mu=1}^{N}\sum_{\nu=1}^{N} E\left\{\phi_{ss}(\Delta_\mu - \Delta_\nu)\right\}, \quad (3.25)$$

but is dependent on the scalar space-continuous autocorrelation function

$$\phi_{ss}(l) = E\left\{s(l_0 + l)s^*(l_0)\right\}. \quad (3.26)$$

The autocorrelation function of a wide-sense stationary random process does not depend on the absolute location l_0 but only on the relative two-dimensional shift l.

For the μ-th displacement error Δ_μ, a 2-D stationary normal distribution with variance σ_Δ^2 and zero mean is assumed where the x- and y-components are statistically independent. The displacement error variance is the same for all N hypotheses. Further, the pairs (Δ_μ, Δ_ν) are assumed to be jointly Gaussian random variables. The correlation coefficient ρ_Δ between two displacement error components $\Delta_{x\mu}$ and $\Delta_{x\nu}$ is the same for all pairs of hypotheses. With these assumptions, the covariance matrix of a displacement error component is given by (3.5). Since the covariance matrix is nonnegative definite, the correlation coefficient ρ_Δ has the limited range according to (3.6).

For the expected values in (3.25), we define a function $g(\omega_0, \sigma_\Delta^2)$ that is only dependent on the spatial correlation of the frame signal $\rho_s = \exp(-\omega_0)$ and the displacement error variance σ_Δ^2. Further, we exploit the fact that the difference of two jointly Gaussian random variables is also Gaussian.

$$E\left\{\phi_{ss}(\Delta_\mu)\right\} = \sigma_s^2 g\left(\omega_0, \sigma_\Delta^2\right) \quad \text{for } \mu = 1, 2, \ldots, N \quad (3.27)$$
$$E\left\{\phi_{ss}(\Delta_\mu - \Delta_\nu)\right\} = \sigma_s^2 g\left(\omega_0, 2\sigma_\Delta^2(1 - \rho_\Delta)\right) \quad \text{for } \mu \neq \nu \quad (3.28)$$

With (3.27) and (3.28), the prediction error variance in (3.25) can be normalized to the frame signal variance σ_s^2. Note that for $\mu = \nu$ the expected value $E\left\{\phi_{ss}(\Delta_\mu - \Delta_\nu)\right\}$ is equal to the variance of the frame signal σ_s^2.

$$\frac{\sigma_e^2}{\sigma_s^2} = \frac{N+1}{N} - 2g\left(\omega_0, \sigma_\Delta^2\right) + \frac{N-1}{N}g\left(\omega_0, 2\sigma_\Delta^2(1 - \rho_\Delta)\right) \quad (3.29)$$

for

$$\frac{1}{1-N} \leq \rho_\Delta \leq 1. \quad (3.30)$$

With the function g in (3.27), we can show that for $N > 1$ the prediction error variance σ_e^2 increases monotonically with the displacement error correlation coefficient ρ_Δ. We know for the single hypothesis predictor that the prediction error variance increases monotonically with the displacement error variance σ_Δ^2. For $N = 1$, (3.29) implies

$$\frac{\partial \sigma_e^2}{\partial \sigma_\Delta^2} = -2\sigma_s^2 \frac{\partial g(\omega_0, \sigma_\Delta^2)}{\partial \sigma_\Delta^2} > 0, \qquad (3.31)$$

that is, the partial derivative of g with respect to σ_Δ^2 is negative. Next, we calculate the partial derivative of the prediction error variance with respect to the displacement error correlation coefficient ρ_Δ for $N > 1$ and observe that this partial derivative is positive.

$$\frac{\partial \sigma_e^2}{\partial \rho_\Delta} = -2\sigma_\Delta^2 \sigma_s^2 \frac{N-1}{N} \frac{\partial g(\omega_0, 2\sigma_\Delta^2(1-\rho_\Delta))}{\partial 2\sigma_\Delta^2(1-\rho_\Delta)} > 0 \qquad (3.32)$$

Consequently, the prediction error variance σ_e^2 increases monotonically with the displacement error correlation coefficient ρ_Δ independent of a particular underlying frame signal model.

3.3 Hypotheses with Additive Noise

To consider signal components that cannot be modeled by motion compensation, statistically independent noise $\mathbf{n}_\mu$ is added to each motion-compensated signal. Further, we assume that the current frame $\mathbf{s}$ originates from a noise-free model video signal $\mathbf{v}$ but is also characterized by statistically independent additive Gaussian noise $\mathbf{n}_0$ [11]. We characterize the residual noise $\mathbf{n}_\mu$ for each hypothesis by the power spectal density $\Phi_{\mathbf{n}_\mu \mathbf{n}_\mu}(\omega)$ and the residual noise $\mathbf{n}_0$ in the current frame by $\Phi_{\mathbf{n}_0 \mathbf{n}_0}(\omega)$. For convenience, we normalize the noise power spectra with respect to the power spectral density of the model video signal $\Phi_{\mathbf{v}\mathbf{v}}(\omega)$.

$$\alpha_\mu(\omega) = \frac{\Phi_{\mathbf{n}_\mu \mathbf{n}_\mu}(\omega)}{\Phi_{\mathbf{v}\mathbf{v}}(\omega)} \qquad \forall \mu = 0, 1, 2, \ldots, N \qquad (3.33)$$

Fig. 3.5 depicts the model for motion-compensated signals with statistically independent additive noise and linear filter. All hypotheses are jointly filtered to determine the final prediction signal. The linear filter is described by the vector valued transfer function $F(\omega)$. In particular, $F(\omega)$ is a row vector with N scalar transfer functions. The power spectrum of the prediction error with the linear filter is

$$\Phi_{ee}(\omega) = \qquad (3.34)$$
$$\Phi_{ss}(\omega) - \Phi_{sc}(\omega) F^H(\omega) - F(\omega) \Phi_{cs}(\omega) + F(\omega) \Phi_{cc}(\omega) F^H(\omega).$$

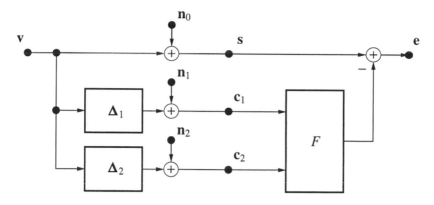

Figure 3.5. Two-hypothesis motion-compensated prediction with linear filter.

In the following, we investigate the performance of motion compensation with complementary hypotheses for both the averaging filter and a Wiener filter.

3.3.1 Averaging Filter

The averaging filter weights each hypothesis equally with the constant factor $\frac{1}{N}$. We use the notation of the column vector $\mathbf{1}$ denoting that all entries are equal to one.

$$F(\omega) = \frac{1}{N}\mathbf{1}^T \tag{3.35}$$

We evaluate the performance of motion compensation with complementary hypotheses for the averaging filter in (3.35) by calculating the rate difference in (2.42). In order to do this, we normalize the power spectral density of the prediction error and substitute the power spectra in (3.34) with (3.3), (3.4), (3.7) and (3.8). Further, we assume individual power spectral densities for the residual noise.

$$\frac{\Phi_{ee}(\omega)}{\Phi_{ss}(\omega)} = \tag{3.36}$$

$$1 + \frac{N + \sum_{\mu=1}^{N}\alpha_{\mu}(\omega)}{N^2(1 + \alpha_0(\omega))} - \frac{2P(\omega, \sigma_\Delta^2)}{1 + \alpha_0(\omega)} + \frac{N-1}{N} \cdot \frac{P\left(\omega, 2\sigma_\Delta^2(1 - \rho_\Delta)\right)}{1 + \alpha_0(\omega)}$$

Experimental results for the predictor design in [15, 16] show that all N hypotheses contribute equally well. Based on this observation, we assume that the noise power spectral densities are identical for all N hypotheses. To simplify the model, we assume also that they are identical to the noise power spectrum of the current frame, i.e., $\alpha_\mu(\omega) = \alpha_0(\omega)$ for $\mu = 1, 2, \ldots, N$. With these assumptions, the normalized prediction error power spectrum for the superimposed predictor with averaging filter reads

$$\frac{\Phi_{ee}(\omega)}{\Phi_{ss}(\omega)} = \frac{N+1}{N} - \frac{2P(\omega, \sigma_\Delta^2)}{1 + \alpha_0(\omega)} + \frac{N-1}{N} \cdot \frac{P\left(\omega, 2\sigma_\Delta^2(1 - \rho_\Delta)\right)}{1 + \alpha_0(\omega)}. \quad (3.37)$$

If there is no noise, i.e., $\alpha_0(\omega) = 0$, we obtain the result in (3.9).

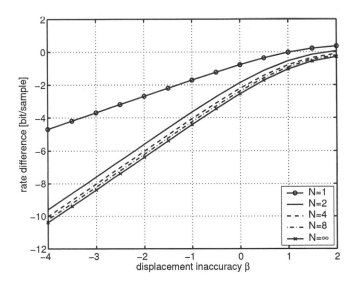

Figure 3.6. Rate difference for motion compensation with complementary hypotheses and averaging filter over the displacement inaccuracy β. Residual noise level RNL = -100 dB.

Figs. 3.6 and 3.7 depict the rate difference for motion compensation with complementary hypotheses over the displacement inaccuracy β at a residual noise level of -100 dB and -30 dB, respectively. The residual noise level is defined by RNL $= 10 \log_{10}(\sigma_n^2)$ where σ_n^2 is the residual noise variance. As suggested in [11], we assume a constant power spectral density for the residual noise. For the plotted range of the motion inaccuracy, a residual noise level of RNL = -100 dB indicates that the residual noise is negligible and the performance is similar to the noiseless case as shown in Fig. 3.4. At a residual noise level of RNL = -30 dB, the rate difference saturates beyond 1/8-th pel accuracy ($\beta = -3$). This is more practical. We observe for motion compensation with complementary hypotheses that the rate difference saturates even at quarter-pel accuracy. Consequently, we can achieve similar prediction performance at lower compensation accuracy when utilizing motion compensation with complementary hypotheses. Regardless of the accuracy of superimposed motion compensation, the rate difference improves for an increasing number of hypotheses due to the noise suppression by the averaging filter.

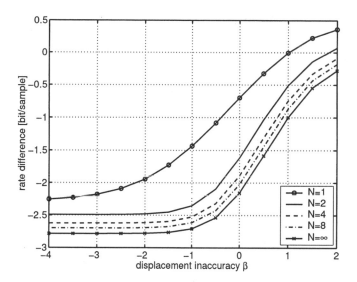

Figure 3.7. Rate difference for motion compensation with complementary hypotheses and averaging filter over the displacement inaccuracy β. Residual noise level RNL = -30 dB.

3.3.2 Wiener Filter

The optimum Wiener filter minimizes the power spectral density of the prediction error. This is a well-known result [205] and we adopt the expression for multihypothesis motion-compensation from [11]. The vector-valued optimum filter $F_o(\omega)$ is the product of the Hermitian conjugate of the cross spectrum, $\Phi_{cs}^H(\omega)$, and the inverse power spectral density matrix of the hypotheses $\Phi_{cc}^{-1}(\omega)$.

$$F_o(\omega) = \Phi_{cs}^H(\omega)\Phi_{cc}^{-1}(\omega) \tag{3.38}$$

This expression minimizes the power spectral density of the prediction error and its minimum is given by

$$\Phi_{ee}(\omega) = \Phi_{ss}(\omega) - \Phi_{cs}^H(\omega)\Phi_{cc}^{-1}(\omega)\Phi_{cs}(\omega). \tag{3.39}$$

To evaluate the performance of the predictor with optimum Wiener filter, we have to specify the vector-valued cross spectrum $\Phi_{cs}(\omega)$ and the power spectral density matrix of the hypotheses $\Phi_{cc}(\omega)$. In the following, we analyze the influence of motion compensation with complementary hypotheses on both the Wiener filter and its prediction performance.

For motion compensation with complementary hypotheses and (3.7), the vector of the cross spectra is simply

$$\Phi_{cs}(\omega) = \mathbf{1}P(\omega, \sigma_\Delta^2)\Phi_{vv}(\omega). \tag{3.40}$$

The cross spectra do not include the power spectral densities of the residual noise as we assume that the individual noise signals are mutually statistically independent. The same argument holds for the non-diagonal entries in the matrix Φ_{cc} which are determined by the characteristic function of the displacement error PDF with the displacement error correlation coefficient ρ_Δ according to (3.8). The diagonal entries in the matrix Φ_{cc} are characterized by the power spectral densities of the hypotheses which include the power spectral densities of the residual noise. We write the matrix Φ_{cc} as the sum of the matrix $\mathbf{11}^T$ and the diagonal matrix diag($\cdot$) as this representation is useful for the following discussion. $\mathbf{11}^T$ is the $N \times N$ matrix with all entries equal to one. With that, the power spectral density matrix for N hypotheses becomes

$$\Phi_{cc}(\omega) = \left[\mathbf{11}^T + \text{diag}\left(\frac{1 + \alpha_i(\omega)}{P_\rho(\omega)} - 1 \right) \right] P_\rho(\omega)\Phi_{vv}(\omega) \qquad (3.41)$$

where $P_\rho(\omega) = P\left(\omega, 2\sigma_\Delta^2(1 - \rho_\Delta)\right)$ abbreviates the characteristic function of the displacement error PDF with the displacement error correlation coefficient ρ_Δ. $\alpha_i(\omega)$ represents the normalized power spectral density of the residual noise in the i-th hypothesis.

The Wiener solution requires the inverse of the power spectral density matrix $\Phi_{cc}(\omega)$. An analytical expression is derived in Appendix A.3 and the optimum Wiener filter according to (3.38) yields

$$F_o(\omega) = \frac{1}{\mathbf{1}^T b(\omega) - 1} \cdot \frac{P(\omega, \sigma_\Delta^2)}{P_\rho(\omega)} b^T(\omega) \qquad (3.42)$$

with the vector

$$b^T(\omega) = \qquad\qquad\qquad\qquad\qquad\qquad\qquad\qquad\qquad (3.43)$$
$$\left(\frac{-P_\rho(\omega)}{1 + \alpha_1(\omega) - P_\rho(\omega)}, \frac{-P_\rho(\omega)}{1 + \alpha_2(\omega) - P_\rho(\omega)}, \ldots, \frac{-P_\rho(\omega)}{1 + \alpha_N(\omega) - P_\rho(\omega)} \right).$$

Please note that only the normalized noise spectrum $\alpha_\mu(\omega)$ differs in the components of $b(\omega)$ and that the contribution from inaccurate motion compensation is the same in each component. The expression for the power spectral density of the prediction error in (3.39) incorporates explicitly the solution for the Wiener filter in (3.38) such that $\Phi_{ee} = \Phi_{ss} - F_o\Phi_{cs}$. After normalization with $\Phi_{ss}(\omega) = (1 + \alpha_0(\omega))\Phi_{vv}(\omega)$, the power spectral density of the prediction error reads

$$\frac{\Phi_{ee}(\omega)}{\Phi_{ss}(\omega)} = 1 - \frac{1}{1 + \alpha_0(\omega)} \cdot \frac{\mathbf{1}^T b(\omega)}{\mathbf{1}^T b(\omega) - 1} \cdot \frac{P^2(\omega, \sigma_\Delta^2)}{P_\rho(\omega)}. \qquad (3.44)$$

As a reminder, $\alpha_0(\omega)$ denotes the normalized power spectral density of the residual noise in the current frame.

If the noise energy in all the hypotheses and the current frame is the same, that is, $\alpha_\mu(\omega) = \alpha(\omega)$ for $\mu = 0, 1, \ldots, N$, we obtain for the optimum Wiener filter

$$F_o(\omega) = \frac{P(\omega, \sigma_\Delta^2)}{1 + \alpha(\omega) + (N-1)P_\rho(\omega)}1^T, \tag{3.45}$$

and for the normalized power spectral density of the prediction error

$$\frac{\Phi_{ee}(\omega)}{\Phi_{ss}(\omega)} = 1 - \frac{1}{1 + \alpha(\omega)} \cdot \frac{NP^2(\omega, \sigma_\Delta^2)}{1 + \alpha(\omega) + (N-1)P_\rho(\omega)}. \tag{3.46}$$

Based on this result, we investigate the influence of motion compensation with complementary hypotheses on both the Wiener filter and its prediction performance.

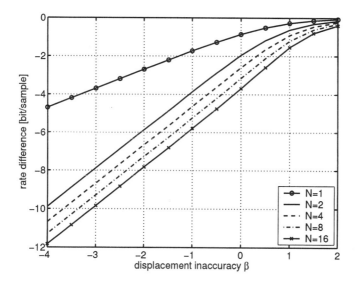

Figure 3.8. Rate difference for motion compensation with complementary hypotheses and Wiener filter over the displacement inaccuracy β. Residual noise level RNL = -100 dB.

Fig. 3.8 depicts the rate difference for motion compensation with complementary hypotheses and Wiener filter over the displacement inaccuracy β and negligible residual noise. For $N > 1$, the graphs show that doubling the number of complementary hypotheses decreases the bit-rate at least by 0.5 bits per sample and the slope reaches up to 2 bits per sample per inaccuracy step. This can also be observed in (3.46) when the residual noise is neglected, i.e. $\alpha(\omega) \to 0$, the displacement error correlation coefficient is set to $\rho_\Delta = \frac{1}{1-N}$, and a Taylor series expansion of second order is applied for the function $P(\omega, \sigma_\Delta^2)$.

$$\frac{\Phi_{ee}(\omega)}{\Phi_{ss}(\omega)} \approx \frac{1}{2}\sigma_\Delta^4(\omega^T\omega)^2 \frac{1}{N-1} \quad \text{for} \quad \sigma_\Delta^2 \to 0 \qquad (3.47)$$

Inserting this result into 2.42 supports the observation in Fig. 3.8 that the slope reaches up to 2 bits per sample per inaccuracy step

$$\Delta R \approx 2\beta + \frac{1}{2}\log_2\left(\frac{1}{N-1}\right) + const. \quad \text{for} \quad \sigma_\Delta^2 \to 0. \qquad (3.48)$$

For very accurate motion compensation, doubling the number of hypotheses results in a rate difference of

$$\Delta R_2 := \Delta R(2N) - \Delta R(N) \approx \frac{1}{2}\log_2\left(\frac{N-1}{2N-1}\right) \quad \text{for} \quad \sigma_\Delta^2 \to 0. \quad (3.49)$$

For a very large number of hypotheses, the rate difference for doubling the number of hypotheses ΔR_2 converges to -0.5 bits per sample. Consequently, the prediction error variance of the optimal noiseless superimposed predictor with Wiener filter converges to zero for an infinite number of hypotheses.

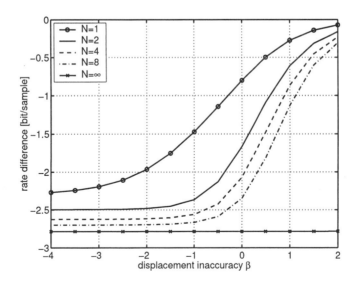

Figure 3.9. Rate difference for motion compensation with complementary hypotheses and Wiener filter over the displacement inaccuracy β. Residual noise level RNL = -30 dB.

Fig. 3.9 shows the rate difference for motion compensation with complementary hypotheses and Wiener filter over the displacement inaccuracy β at a residual noise level of -30 dB. Similar to the averaging filter for motion compensation with complementary hypotheses in Fig. 3.7, the rate difference saturates even at for quarter-pel accuracy. And for very accurate superimposed motion compensation, the rate difference improves for an increasing number of hypotheses due to the noise suppression by the Wiener filter but saturates for $N \rightarrow \infty$. In contrast to the averaging filter, the Wiener filter with a very large number of hypotheses is able to eliminate the influence of motion inaccuracy. This can be observed in (3.46) for $N \rightarrow \infty$. In this case, the normalized power spectral density of the prediction error yields

$$\frac{\Phi_{ee}(\omega)}{\Phi_{ss}(\omega)} = 1 - \frac{1}{1 + \alpha(\omega)} \quad \text{for} \quad N \rightarrow \infty. \tag{3.50}$$

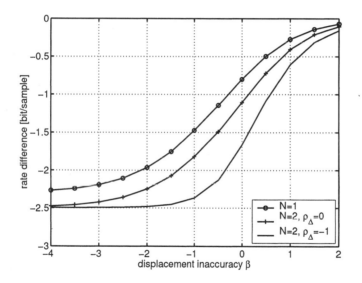

Figure 3.10.   Rate difference for superimposed motion-compensated prediction with Wiener filter over the displacement inaccuracy β for both optimized displacement error correlation and statistically independent displacement error. Residual noise level RNL = -30 dB. In all cases, the optimum filter is applied.

Fig. 3.10 depicts the rate difference for superimposed motion-compensated prediction over the displacement inaccuracy β for both optimized displacement error correlation and statistically independent displacement error. The residual noise level is chosen to be -30 dB. For half-pel accurate motion compensation and 2 hypotheses, we gain about 0.6 bits/sample in rate difference for motion compensation with complementary hypotheses over statistically independent displacement error. This corresponds to a prediction gain of about 3.6 dB.

So far, we investigated the prediction performance achieved by the optimum Wiener filter for motion compensation with complementary hypotheses. In the following, we discuss the transfer function of the optimum Wiener filter according to (3.45) for motion compensation with $N = 2$ complementary hypotheses and compare it to that of the optimum Wiener filter for motion compensation with $N = 1$ hypothesis.

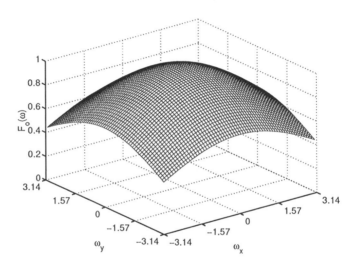

Figure 3.11.   Transfer function of the Wiener filter for $N = 1$, $\beta = 0$, and no residual noise.

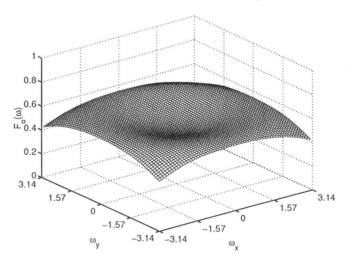

Figure 3.12.   Transfer function of a component of the Wiener filter for $N = 2$, $\beta = 0$, and no residual noise.

Figs. 3.11 and 3.12 depict the transfer function of one component of the optimum Wiener filter according to (3.45) for $N = 1$ and $N = 2$, respectively. In both cases, we neglect the residual noise and use integer-pel accurate motion compensation. The transfer function of the filter for the superimposed predictor is flatter and, hence, shows significantly less spectral selectivity when compared to the case of the single hypothesis predictor. In addition, the transfer function for motion compensation with complementary hypotheses seems to suppress low frequency components. We will investigate this further by means of cross sections of the transfer functions.

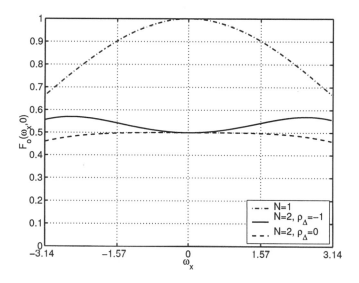

Figure 3.13. Cross section at $\omega_y = 0$ of the transfer function of the Wiener filter for $\beta = 0$ and no residual noise. The one-hypothesis filter $N = 1$ is compared to the two-hypothesis filter $N = 2$ with optimized displacement error correlation coefficient $\rho_\Delta = -1$ and uncorrelated displacement error $\rho_\Delta = 0$.

Figs. 3.13 and 3.14 show cross sections at $\omega_y = 0$ of the transfer functions of the optimum Wiener filter according to (3.45) for $\beta = 0$ and $\beta = 1$, respectively. The one-hypothesis filter $N = 1$ is compared to the two-hypothesis filter $N = 2$ with optimized displacement error correlation coefficient $\rho_\Delta = -1$ and uncorrelated displacement error $\rho_\Delta = 0$. The spectral selectivity decreases for more accurate motion compensation. This allows us to accurately compensate high frequency components. Interestingly, the filter for motion compensation with complementary hypotheses amplifies high frequency components when compared to the two-hypothesis filter with uncorrelated displacement error. In other words, the optimum Wiener filter for motion compensation with complementary hypotheses promotes high frequency components that are otherwise

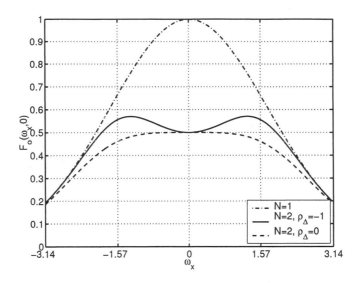

Figure 3.14. Cross section at $\omega_y = 0$ of the transfer function of the Wiener filter for $\beta = 1$ and no residual noise. The one-hypothesis filter $N = 1$ is compared to the two-hypothesis filter $N = 2$ with optimized displacement error correlation coefficient $\rho_\Delta = -1$ and uncorrelated displacement error $\rho_\Delta = 0$.

suppressed by the two-hypothesis filter with uncorrelated displacement error. These effects grow larger for less accurate motion compensation.

As previously mentioned, very accurate motion compensation flattens the characteristic function of the displacement error PDF, that is, $P(\omega, \sigma^2) = 1$ for $\sigma_\Delta^2 \to 0$. In this case, the optimum Wiener filter in (3.45) depends mainly on the power spectral density of the residual noise according to

$$F_o(\omega) = \frac{1}{N + \alpha(\omega)} \mathbf{1}^T. \qquad (3.51)$$

If we neglect the residual noise in (3.51), i.e., $\alpha(\omega) \to 0$, we obtain the averaging filter according to (3.35). In other words, considering only signal components that capture the motion in the model, the averaging filter is, in the limit, the optimum filter for very accurate motion compensation. Note, that the complementary hypotheses are not identical, even for very accurate motion compensation.

3.4   Forward-Adaptive Hypothesis Switching

Superimposed motion-compensated prediction combines more than one motion-compensated signal, or hypothesis, to predict the current frame signal. In particular, we do not specify how the multiple hypotheses are selected. Now, we assume that they are determined by forward-adaptive hypothesis switching [206]. Hypothesis switching selects one motion-compensated signal from a set of M reference frames. The parameter which indicates a particular frame in the set of M reference frames is transmitted as side information to the decoder. This models multiframe motion-compensated prediction as discussed in Section 2.2.4. In the following, we discuss this concept of forward-adaptive hypothesis switching for superimposed motion-compensated prediction and its performance with complementary hypotheses.

Assume that we linearly combine N hypotheses. Each hypothesis that is used for the combination is selected from a set of motion-compensated signals of size M. We study the influence of the hypothesis set size M on both the accuracy of motion compensation of forward-adaptive hypothesis switching and the efficiency of superimposed motion estimation. In both cases, we examine the noise-free limiting case. That is, we neglect signal components that are not predictable by motion compensation. Selecting one hypothesis from a set of motion-compensated signals of size M, that is, switching among M hypotheses, reduces the displacement error variance by factor M, if we assume statistically independent displacement errors. Integrating forward-adaptive hypothesis switching into superimposed motion-compensated prediction, that is, allowing a combination of switched hypotheses, increases the gain of superimposed motion-compensated prediction over the single hypothesis case for growing hypothesis set size M.

Experimental results in Section 4.3.3 and 5.3.2 confirm that block-based multiframe motion compensation enhances the efficiency of superimposed prediction. The experimental setup is such that we combine both block-based predictors and superimpose N hypotheses where each hypothesis is obtained by switching among M motion-compensated blocks.

3.4.1 Signal Model for Hypothesis Switching

A signal model for hypothesis switching is depicted in Fig. 3.15 for two hypotheses. The current frame signal $s[l]$ at discrete location $l = (x, y)$ is predicted by selecting between M hypotheses $c_\mu[l]$ with $\mu = 1, ..., M$. The resulting prediction error is denoted by $e[l]$.

To capture the limited accuracy of motion compensation, we associate a vector valued displacement error Δ_μ with the μ-th hypothesis c_μ. The displacement error reflects the inaccuracy of the displacement vector used for motion compensation. We assume a 2-D stationary normal distribution with variance σ_Δ^2 and zero mean where x- and y-components are statistically independent. The displacement error variance is the same for all M hypotheses.

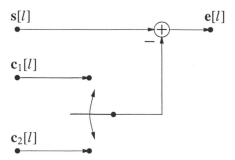

Figure 3.15.   Forward-adaptive hypothesis switching for motion-compensated prediction.

This is reasonable because all hypotheses are compensated with the same accuracy. Further, the pairs (Δ_μ, Δ_ν) are assumed to be statistically independent Gaussian random variables and that the displacement error for all hypotheses are spatially constant.

For simplicity, we assume that all hypotheses c_μ are shifted versions of the current frame signal s. The shift is determined by the displacement error Δ_μ of the μ-th hypothesis. For that, the ideal reconstruction of the space-discrete signal $s[l]$ is shifted by the continuous valued displacement error and re-sampled on the original orthogonal grid.

This model neglects "noisy" signal components and assumes that motion accuracy is basically the decision criterion for switching. As previously discussed in Section 3.2.3, prediction error variance decreases by reducing the displacement error variance of hypotheses that are used for prediction. But a smaller displacement error variance can only be achieved by increasing the probability of individual displacement error that are "close" to the origin. In other words, we select from the set of M motion-compensated signals the one with the smallest displacement error.

3.4.2   Minimizing the Radial Displacement Error

Hypothesis switching improves the accuracy of motion compensation by selecting among M hypotheses the one with the smallest displacement error. Now, let us assume that the components of the displacement error for each hypothesis are i.i.d. Gaussian. The Euclidean distance to the zero displacement error vector defines the *radial displacement error* for each hypothesis.

$$\Delta_{r\mu} = \sqrt{\Delta_{x_\mu}^2 + \Delta_{y_\mu}^2} \tag{3.52}$$

We assume that the hypothesis with minimum radial displacement error

$$\Delta_r^M = \min_{\mu}(\Delta_{r1}, \ldots, \Delta_{r\mu}, \ldots, \Delta_{rM}) \tag{3.53}$$

is used to predict the signal.

In the following, we describe hypothesis switching by means of the reliability function [205] of the minimum radial displacement error. The reliability function $R_{\mathbf{\Delta_r}^M}(r)$ is closely related to the distribution function and is defined as the probability of the event that $\mathbf{\Delta_r}^M$ is larger than r.

$$R_{\mathbf{\Delta_r}^M}(r) = \Pr\{\mathbf{\Delta_r}^M > r\} \qquad (3.54)$$

The reliability function of the minimum radial displacement error can be expressed in terms of the reliability function of the set of M hypotheses. The probability of the event that the minimum radial displacement error is larger than r is equal to the probability of the event that all radial displacement errors are larger than r.

$$\begin{aligned}
R_{\mathbf{\Delta_r}^M}(r) &= \Pr\{\min_{\mu}(\mathbf{\Delta}_{r1}, \ldots, \mathbf{\Delta}_{r\mu}, \ldots, \mathbf{\Delta}_{rM}) > r\} \\
&= \Pr\{\mathbf{\Delta}_{r1} > r, \ldots, \mathbf{\Delta}_{rM} > r\} \\
&= R_{\mathbf{\Delta}_{r1}\ldots\mathbf{\Delta}_{rM}}(r, \ldots, r) \qquad (3.55)
\end{aligned}$$

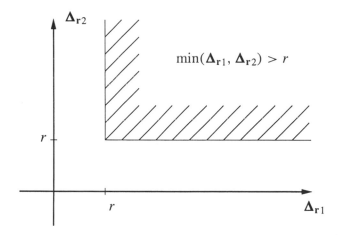

Figure 3.16. The area in which the minimum of $\mathbf{\Delta}_{r1}$ and $\mathbf{\Delta}_{r2}$ is larger than r.

Fig. 3.16 depicts an example for switching two hypotheses. It marks the area in which the minimum of two radial displacement errors is larger than r. Consequently, the probability of the event that the minimum radial displacement error is larger than r is equal to the probability of the event that both radial displacement errors are larger than r.

Each displacement error is drawn from a 2-D normal distribution with zero mean and variance $\sigma_{\mathbf{\Delta}_\mu}^2$. The displacement errors of the M hypotheses are assumed to be statistically independent. The x- and y-components of the displacement errors are arranged to vectors $\mathbf{\Delta_x}$ and $\mathbf{\Delta_y}$, respectively.

$$p_{\mathbf{\Delta_x \Delta_y}}(\Delta_x, \Delta_y) = \frac{1}{(2\pi)^M \det(C_{\mathbf{\Delta_x \Delta_x}})} e^{-\frac{1}{2}\left[\Delta_x^T C_{\mathbf{\Delta_x \Delta_x}}^{-1}\Delta_x + \Delta_y^T C_{\mathbf{\Delta_x \Delta_x}}^{-1}\Delta_y\right]} \quad (3.56)$$

with

$$C_{\mathbf{\Delta_x \Delta_x}} = C_{\mathbf{\Delta_y \Delta_y}} = \begin{pmatrix} \sigma_{\Delta_{x1}}^2 & 0 & \cdots & 0 \\ 0 & \sigma_{\Delta_{x2}}^2 & \cdots & 0 \\ \vdots & \vdots & \ddots & \vdots \\ 0 & 0 & \cdots & \sigma_{\Delta_{xM}}^2 \end{pmatrix} \quad (3.57)$$

The criterion for switching is the radial displacement error. To obtain a closed-form expression of the radial displacement error PDF, it is assumed that the variances in x- and y-direction are identical. This is reasonable if the accuracy of motion compensation is identical for both dimensions. With this assumption, we can easily determine the probability density function of $\mathbf{\Delta_r}$ [205]:

$$p_{\mathbf{\Delta_r}}(\Delta_r) = \left(\prod_{\mu=1}^M \Delta_{r\mu}\right) \frac{1}{\det(C_{\mathbf{\Delta_x \Delta_x}})} e^{-\frac{1}{2}\Delta_r^T C_{\mathbf{\Delta_x \Delta_x}}^{-1}\Delta_r} \quad (3.58)$$

An M-dimensional Rayleigh PDF is obtained describing M independent radial displacement errors.

In order to minimize the radial displacement error, the M-dimensional reliability function of the displacement error has to be determined.

$$\begin{aligned} R_{\mathbf{\Delta_{r1} \cdots \Delta_{rM}}}(\Delta_{r1} \cdots \Delta_{rM}) &= \int_{\Delta_{r1}}^{\infty} \cdots \int_{\Delta_{rM}}^{\infty} p_{\mathbf{\Delta_r}}(u)\,du \\ &= e^{-\frac{1}{2}\Delta_r^T C_{\mathbf{\Delta_x \Delta_x}}^{-1}\Delta_r} \end{aligned} \quad (3.59)$$

The reliability function of the minimum radial displacement error $R_{\mathbf{\Delta_r}M}(r)$ is obtained by evaluating the M-dimensional reliability function at the same value r for all dimensions: $R_{\mathbf{\Delta_r}M}(r) = R_{\mathbf{\Delta_{r1} \cdots \Delta_{rM}}}(\mathbf{1}r)$. $\mathbf{1}$ denotes the vector with all components equal to one. The minimum radial displacement error is also Rayleigh distributed. Note that a one-dimensional PDF is given by the negative derivative of the reliability function.

$$R_{\mathbf{\Delta_r}M}(r) = e^{-\frac{1}{2}\frac{r^2}{\tau^2}} \quad (3.60)$$

$$p_{\mathbf{\Delta_r}M}(r) = \frac{r}{\tau^2} e^{-\frac{1}{2}\frac{r^2}{\tau^2}} \quad (3.61)$$

$$\text{with} \quad \tau^2 = \frac{1}{\mathbf{1}^T C_{\mathbf{\Delta_x \Delta_x}}^{-1}\mathbf{1}} \quad (3.62)$$

The variance of the minimum radial displacement error is of interest. The covariance matrix of the Rayleigh PDF in (3.58) is $C_{\Delta_r \Delta_r} = (2 - \frac{\pi}{2}) C_{\Delta_x \Delta_x}$ and the variance of the switched radial displacement error is given by $\sigma^2_{\Delta_r M} = (2 - \frac{\pi}{2}) \tau^2$ [140]. In order to omit the constant factor, the variance of the minimum radial displacement error is stated as a function of the covariance matrix $C_{\Delta_r \Delta_r}$.

$$\sigma^2_{\Delta_r M} = \frac{1}{1^T C_{\Delta_r \Delta_r}^{-1} 1} \tag{3.63}$$

For example, the variances of the radial displacement errors might be identical for all M hypotheses. (3.63) implies that switching of independent Rayleigh distributed radial displacement errors reduces the variance by factor M.

$$\sigma^2_{\Delta_r M} = \frac{\sigma^2_{\Delta_r}}{M}. \tag{3.64}$$

3.4.3 Equivalent Predictor

Section 3.4.2 shows that both the individual radial displacement errors and the minimum radial displacement error are Rayleigh distributed. This suggests to define an equivalent motion-compensated predictor for switched prediction that uses just one hypothesis but with a much smaller the displacement error variance. The equivalent distribution of the displacement error is assumed to be separable and normal with zero mean and variance

$$\sigma^2_{\Delta_x M} = \frac{1}{1^T C_{\Delta_x \Delta_x}^{-1} 1}. \tag{3.65}$$

Figure 3.17 gives an example for the equivalent predictor when switching M independent and identically distributed displacement error signals. The 2-D and Gaussian distributed displacement error of each hypothesis is transformed into the Rayleigh distributed radial displacement error. Minimizing M independent radial displacement error signals yields a radial displacement error which is also Rayleigh distributed. The inverse transform into the 2-D Cartesian coordinate system determines the 2-D normal distributed displacement error of the equivalent predictor. Its displacement error variance is reduced by factor M in comparison to the variance of individual hypotheses.

The equivalent predictor suggests a model for switched prediction with just one hypothesis but reduced displacement error variance. This reduced variance determines the accuracy of motion-compensated prediction with forward-adaptive hypothesis switching. That is, switching improves the accuracy of motion-compensated prediction by reducing the displacement error variance.

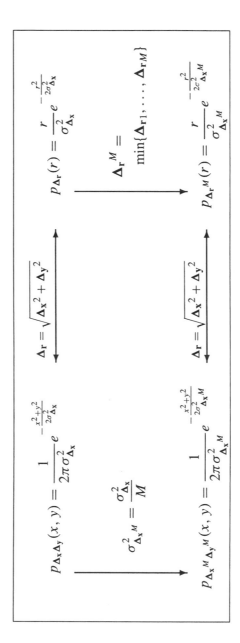

$$p_{\Delta_x \Delta_y}(x, y) = \frac{1}{2\pi \sigma^2_{\Delta_x}} e^{-\frac{x^2+y^2}{2\sigma^2_{\Delta_x}}}$$

$$\Delta_r = \sqrt{\Delta_x^2 + \Delta_y^2}$$

$$p_{\Delta_r}(r) = \frac{r}{\sigma^2_{\Delta_x}} e^{-\frac{r^2}{2\sigma^2_{\Delta_x}}}$$

$$\sigma^2_{\Delta_x^M} = \frac{\sigma^2_{\Delta_x}}{M}$$

$$\Delta_r^M = \min\{\Delta_{r1}, \ldots, \Delta_{rM}\}$$

$$p_{\Delta_x^M \Delta_y^M}(x, y) = \frac{1}{2\pi \sigma^2_{\Delta_x^M}} e^{-\frac{x^2+y^2}{2\sigma^2_{\Delta_x^M}}}$$

$$\Delta_r = \sqrt{\Delta_x^2 + \Delta_y^2}$$

$$p_{\Delta_r^M}(r) = \frac{r}{\sigma^2_{\Delta_x^M}} e^{-\frac{r^2}{2\sigma^2_{\Delta_x^M}}}$$

Figure 3.17. The probability density functions of Cartesian and radial displacement errors for switching M independent and identically distributed displacement error.

3.4.4 Motion Compensation with Complementary Hypotheses and Forward-Adaptive Hypothesis Switching

We combine forward-adaptive hypothesis switching with complementary hypotheses motion-compensated prediction such that we superimpose N complementary hypotheses where each hypothesis is obtained by switching among M motion-compensated signals.

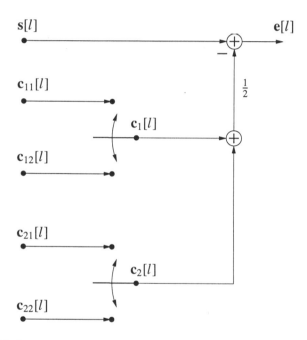

Figure 3.18. Motion-compensated prediction with $N = 2$ complementary hypotheses and forward-adaptive hypothesis switching $M = 2$.

Fig. 3.18 depicts a prediction scheme where the complementary hypotheses c_1 and c_2 are determined by switching for each hypothesis among $M = 2$ motion-compensated signals. In practice, $N = 2$ complementary hypotheses are selected from $M = 2$ reference frames. Further, as the complementary hypotheses are jointly estimated, the frame selection is also performed jointly.

Let us consider the example that we switch among M motion-compensated signals and utilize complementary hypotheses motion compensation with $N = 2$ hypotheses c_1 and c_2. Then, complementary hypotheses motion compensation uses two sets of motion-compensated signals of size M. The hypothesis c_1 is selected from the set $\{c_{11}, \ldots, c_{1M}\}$, and the complementary hypothesis c_2 from the complementary set $\{c_{21}, \ldots, c_{2M}\}$. Choosing one motion-compensated signal from each set provides two hypotheses whose displacement error correlation coefficient is -1. But choosing two motion-compensated signals from the same set provides two signals whose displacement error correlation coefficient is 0. For this example, we assume that these hypotheses exist and that an ideal superimposed motion estimator is able to determine the desired signals.

According to the previous section, choosing among M motion-compensated signals can reduce the displacement error variance by up to a factor of M. Motion compensation with complementary hypotheses utilizes for each hypothesis forward-adaptive switching. Consequently, the displacement error variance of superimposed hypotheses σ_Δ^2 in (3.9) is smaller by a factor of M

$$\frac{\Phi_{ee}(\omega)}{\Phi_{ss}(\omega)} = \frac{N+1}{N} - 2P\left(\omega, \frac{\sigma_\Delta^2}{M}\right) + \frac{N-1}{N}P\left(\omega, 2\frac{\sigma_\Delta^2}{M}(1-\rho_\Delta)\right). \quad (3.66)$$

Note, that (3.66) represents a performance bound given the previous assumptions.

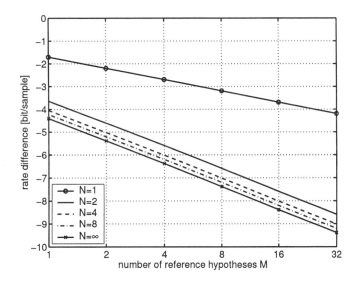

Figure 3.19.   Rate difference over the number of motion-compensated signals M for motion compensation with complementary hypotheses and forward-adaptive hypothesis switching. The switched hypotheses are just averaged and no residual noise is assumed. The results are for half-pel accurate motion compensation, i.e., $\beta = -1$.

Fig. 3.19 depicts the rate difference over the size of the motion-compensated signal set M according to (3.66). The performance bound of motion compensation with complementary hypotheses and forward-adaptive hypothesis switching for $N = 2, 4, 8$, and ∞ linearly combined hypotheses is compared to motion-compensated prediction with forward-adaptive hypothesis switching ($N = 1$). Half-pel accurate motion compensation ($\beta = -1$) is assumed. We observe that doubling the number of reference hypotheses decreases the bit-rate for motion-compensated prediction by 0.5 bits per sample and for motion-compensated prediction with complementary hypotheses by 1 bit per sample. Due to the different slopes, the bit-rate savings by complementary hypotheses motion compensation over single hypothesis motion compensation increase with the number of available motion-compensated signals M. This theoretical finding supports the experimental results in Section 4.3.3 and 5.3.2.

The experimental results show also a saturation of the gain by forward-adaptive hypothesis switching. Choosing among M motion-compensated signals can reduce the displacement error variance by up to a factor of M. This lower bound is obtained when switching among M motion-compensated signals with statistically independent displacement error. Correlated displacement error degrade the performance of hypothesis switching and cause a saturation of the bit-rate savings for an increasing number of motion-compensated signals M.

3.5   Pictures with Varying Number of Hypotheses

In practice, video sequences are usually coded with two different picture types: P- and B-pictures. P-pictures use block-based (multiframe) motion-compensated prediction with temporally prior reference pictures whereas B-pictures utilize block-based bidirectional prediction with one (multiple) temporally subsequent and one (multiple) temporally prior reference picture(s). Picture type sequences with a varying number of inserted B-pictures like P, PB, PBB, and PBBB are widely used. Bidirectional prediction in B-pictures is a special case of superimposed motion-compensated prediction with $N = 2$ hypotheses. If we neglect OBMC, motion-compensated prediction in P-pictures is predominantly single hypothesis prediction.

In the following, we investigate the impact of motion compensation with complementary hypotheses on the sequence coding performance based on the previously discussed high-rate approximation for sequential encoding. At high rates, the residual encoder guarantees infinitesimal small reconstruction distortion and any picture coding order that exploits all statistical dependencies provides the same coding efficiency. B-pictures with complementary hypothesis motion compensation provide improved prediction performance over P-pictures and their relative occurrence improves sequence coding performance. Because of the high-rate approximation, we neglect the rate of the side information.

The average rate difference for predictive sequence coding compared to optimum intra-frame sequence coding yields

$$\Delta R = \frac{n_P}{n_P + n_B} \Delta R_P + \frac{n_B}{n_P + n_B} \Delta R_B, \tag{3.67}$$

where n_P and n_B are the number of P- and B-pictures in the sequence. ΔR_P denotes the rate difference of single hypothesis prediction and ΔR_B that of motion compensation with complementary hypotheses. The relative number of P- and B-pictures in a sequence determines the contribution of a particular picture type to the average rate difference.

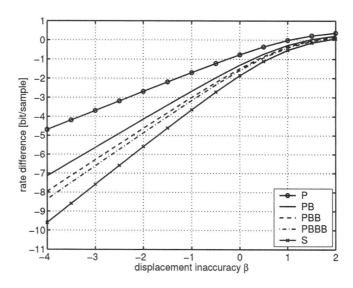

Figure 3.20. Performance bounds for five different picture type sequences. The rate difference is depicted over the displacement inaccuracy. B-pictures average two complementary hypotheses. No residual noise is assumed.

Fig. 3.20 depicts the rate difference for predictive sequence coding compared to optimum intra-frame sequence coding over the motion accuracy β. No residual noise is assumed. The rate difference is given for five different picture type sequences: P, PB, PBB, and PBBB with 0, 6, 8, and 9 B-pictures in a group of 12 pictures where the B-pictures utilize motion-compensated prediction with two complementary hypotheses. S denotes a sequence of superimposed pictures which also use motion-compensated prediction with two complementary hypotheses. As the slope of the rate difference is not equal for P- and B-pictures, the linear combination in (3.67) affects the slope of the average rate difference. In the limit, the picture type sequences PB, PBB, and PBBB modify the slope of the average rate difference to 3/2, 5/3, and 7/4 bits per sample per inaccuracy step, respectively.

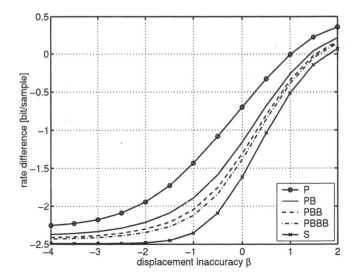

Figure 3.21. Performance bounds for five different picture type sequences. The rate difference is depicted over the displacement inaccuracy. B-pictures average two complementary hypotheses. The residual noise is -30 dB.

Fig. 3.21 depicts similar to Fig. 3.20 the rate difference for predictive sequence coding compared to optimum intra-frame sequence coding over the motion accuracy β but with a residual noise of RNL = -30 dB. In the presence of residual noise, the advantage due to the slope difference is mainly prevalent in the range between quarter-pel and integer-pel accuracy. Beyond quarter-pel accuracy, the residual noise dominates and the average rate difference starts to saturate.

3.6 Conclusions

This chapter extends the theory of multihypothesis motion-compensated prediction by introducing the concept of motion-compensation with complementary hypotheses. We allow for the displacement errors of N hypotheses to be correlated. The assumption that the N displacement errors are jointly distributed imposes an constraint on the displacement error correlation coefficient as any covariance matrix is nonnegative definite. We analyze the dependency between the displacement error correlation coefficient and the performance of superimposed motion compensation. A ideal superimposed motion estimator minimizes the prediction error and, consequently, the displacement error correlation coefficient. The optimal ideal superimposed motion estimator is the estimator that determines N complementary hypotheses with maximally negatively correlated displacement error. This is a result which is independent of the underlying frame signal model.

Assuming band-limited and noise-free frame signals, the motion compensator with complementary hypotheses achieves a rate difference of up to 2 bits per sample per inaccuracy step whereas a single hypothesis motion compensator is limited to 1 bit per sample per inaccuracy step. When averaging more than two hypotheses, the bit-rate savings are limited even if the number of hypotheses grows very large. When utilizing the optimum Wiener filter, the bit-rate savings are not limited and doubling the number of hypotheses improves the rate difference by 0.5 bits per sample for a very large number of hypotheses.

In the presence of residual noise, the bit-rate savings saturate for increasing motion accuracy. The optimum Wiener filter with an infinite number of hypotheses permits bit-rate savings that are independent of the motion accuracy and limited only by the residual noise. In addition, the Wiener filter for motion compensation with complementary hypotheses amplifies high frequency components and shows band-pass characteristics. The optimum Wiener filter for single hypothesis motion compensation shows only low-pass characteristics.

This chapter also combines forward-adaptive hypothesis switching with complementary hypothesis motion compensation. Choosing among M motion-compensated signals with statistically independent displacement error reduces the displacement error variance by up to a factor of M. We utilize motion compensation with complementary hypotheses such that each hypothesis is determined by switching among M reference hypotheses. An analysis of the noise-free case shows that doubling the number of reference hypotheses decreases the bit-rate of motion-compensated prediction by 0.5 bits per sample and that of motion compensation with complementary hypotheses by 1 bit per sample. Due to the different slopes, the bit-rate savings by complementary hypotheses motion compensation over single hypothesis motion compensation increase with the number of available reference signals M.

Finally, this chapter discusses sequence coding with different picture types characterized by a varying number of hypotheses. As the slope of the rate difference is not equal for single hypothesis prediction (P-pictures) and superimposed prediction with complementary hypotheses (B-pictures), particular picture type sequences influence the slope of the overall average rate difference. The analysis suggests that the number of pictures with complementary hypotheses motion compensation should be increased such that the overall average rate difference benefits from complementary hypotheses.

Chapter 4

ITU-T REC. H.263 AND
SUPERIMPOSED PREDICTION

4.1 Introduction

This chapter discusses a practical implementation of superimposed motion-compensated prediction according to Section 2.2.5, based on the ITU-T Recommendation H.263 [63, 14]. The general concept of superimposed prediction with multiple reference frames is not part of this standard. We investigate the efficient number of hypotheses, the combination with variable block size prediction, and the influence of multiframe motion compensation. Further, we relate the experimental results to the insights from Chapter 3.

ITU-T Recommendation H.263 utilizes a hybrid video coding concept with block-based motion-compensated prediction and DCT-based transform coding of the prediction error. P-frame coding of H.263 employs INTRA and INTER coding modes. Superimposed motion-compensated prediction for P-frame coding is enabled by new coding modes that are derived from H.263 INTER coding modes. Annex U of ITU-T Rec. H.263 allows multiframe motion-compensated prediction but does not provide capabilities for superimposed prediction. A combination of H.263 Annex U with B-frames leads to the concept of superimposed multiframe prediction. In this chapter, we do not use H.263 B-frames as we discuss interpolative prediction for in-order encoding of sequences. H.263 B-frames can only be used for out-of-order encoding of sequences. Further, the presented concept of superimposed multiframe prediction is much more general than the B-frames in H.263. ITU-T Rec. H.263 also provides OBMC capability. As discussed previously, OBMC uses more than one motion vector for predicting the same pixel but those motion vectors are also used by neighboring blocks. In this chapter, a block predicted by superimposed motion compensation has its individual set of motion vectors. We do not overlap shifted blocks that might be obtained by utilizing spatially neighboring motion vectors. The INTER4V coding mode of H.263 utilizes VBS prediction with either OBMC or an in-loop deblocking filter. Superimposed prediction already filters the motion-compensated signals and an extension of H.263 OBMC or in-loop deblocking filter is not implemented.

The outline of this chapter is as follows: Section 4.2 explains the video codec with superimposed motion-compensated prediction. We outline syntax extensions for H.263 as well as coder control issues for mode decision and motion estimation. Section 4.3 discusses several experiments. We investigate the efficient number of hypotheses, combine superimposed prediction with blocks of variable size, and study the influence of multiframe motion compensation.

4.2   Video Coding with Superimposed Motion

ITU-T Recommendation H.263 standardizes a block-based hybrid video codec. Such a codec utilizes motion-compensated prediction to generate a prediction signal from previous reconstructed frames in order to reduce the bit-rate of the residual encoder. For block-based MCP, one motion vector and one picture reference parameter which address the reference block in a previous reconstructed frame are assigned to each block in the current frame.

The superposition video codec [207] additionally reduces the bit-rate of the residual encoder by improving the prediction signal. The improvement is achieved by combining linearly more than one motion-compensated prediction signal. For block-based superimposed MCP, more than one motion vector and picture reference parameter, which address a reference block in previous reconstructed frames, is assigned to each block in the current frame. These multiple reference blocks are linearly combined to form the block-based superimposed prediction signal.

The coding efficiency is improved at the expense of increased computational complexity for motion estimation at the encoder. But this disadvantage can be tackled by efficient estimation strategies like successive elimination as discussed in Section 2.3.4. At the decoder, a minor complexity increase is caused by the selection and combination of multiple prediction signals. Please note that not all macroblocks utilize superimposed MCP.

4.2.1   Syntax Extensions

The syntax of H.263 is extended such that superimposed motion compensation is possible. On the macroblock level, two new modes, INTER2H and INTER4H, are added which allow two or four hypotheses per macroblock, respectively. These modes are similar to the INTER mode of H.263. The IN-TER2H mode additionally includes an extra motion vector and frame reference parameter for the second hypothesis. The INTER4H mode incorporates three extra motion vectors and frame reference parameters. For variable block size prediction, the INTER4V mode of H.263 is extended by a superposition block pattern. This pattern indicates for each 8×8 block the number of motion vectors and frame reference parameters. This mode is called INTER4VMH. The superposition block pattern has the advantage that the number of hypotheses can be indicated individually for each 8×8 block. This allows the important case that just one 8×8 block can be coded with more than one motion vector and frame reference parameter. The INTER4VMH mode includes the INTER4V mode when the superposition block pattern indicates just one hypothesis for all 8×8 blocks.

4.2.2 Coder Control

The coder control for the superposition video codec utilizes rate-distortion optimization by Lagrangian methods. For that, the average Lagrangian cost of a macroblock, given the previously encoded macroblocks, is minimized. The average cost $J = D + \lambda R$ consists of the average distortion D and the weighted average bit-rate R. The weight, also called Lagrange multiplier λ, is related to the macroblock quantization parameter QP by the relationship

$$\lambda = 0.85 QP^2 \qquad (4.1)$$

as discussed in Section 2.3.3. This generic optimization method provides the encoding strategy for the superposition encoder: Minimizing the instantaneous Lagrangian costs for each macroblock minimizes the average Lagrangian costs, given the previous encoded macroblocks.

H.263 allows several encoding modes for each macroblock. The one with the lowest Lagrangian costs is selected for the encoding. This strategy is also called rate-constrained mode decision [208], [121].

The new superposition modes include both superimposed prediction and prediction error encoding. The Lagrangian costs of the new superposition modes have to be evaluated for rate-constrained mode decision. The distortion of the reconstructed macroblock is determined by the summed squared error. The macroblock bit-rate includes also the rate of all motion vectors and picture reference parameters. This allows the best trade-off between superimposed MCP rate and prediction error rate [122].

As already mentioned, superimposed MCP improves the prediction signal by spending more bits for the side-information associated with the motion-compensating predictor. But the encoding of the prediction error and its associated bit-rate also determines the quality of the reconstructed block. A joint optimization of superimposed motion estimation and prediction error encoding is far too demanding. But superimposed motion estimation independent of prediction error encoding is an efficient and practical solution. This solution is efficient if rate-constrained superimposed motion estimation, as explained before, is applied.

For example, the encoding strategies for the INTER and INTER2H modes are as follows: Testing the INTER mode, the encoder performs successively rate-constrained motion estimation for integer-pel positions and rate-constrained half-pel refinement. Rate-constrained motion estimation incorporates the prediction error of the video signal as well as the bit-rate for the motion vector and picture reference parameter. Testing the INTER2H mode, the encoder performs rate-constrained superimposed motion estimation. Rate-constrained superimposed motion estimation incorporates the superimposed prediction error of the video signal as well as the bit-rate for two motion vectors and picture reference parameters. Rate-constrained superimposed motion estimation is performed by the HSA in Fig. 2.6 which utilizes in each iteration step rate-constrained motion estimation to determine a conditional rate-constrained motion estimate. Given the obtained motion vectors and picture reference parameters for the INTER and INTER2H modes, the resulting prediction errors are encoded to evaluate the mode costs. The encoding strategy for the INTER4H mode is similar. For the INTER4VMH mode, the number of hypotheses for each 8×8 block is determined after encoding its residual error.

4.3 Experimental Results

The superposition codec is based on the ITU-T Rec. H.263 with unrestricted motion vector mode, four motion vectors per macroblock, and enhanced reference picture selection in sliding window buffering mode. In contrast to H.263, the superposition codec uses a joint entropy code for horizontal and vertical motion vector data as well as an entropy code for the picture reference parameter. The efficiency of the reference codec is comparable to those of the H.263 test model TMN-10 [209]. The test sequences are coded at QCIF resolution and 10 fps. Each sequence has a length of ten seconds. For comparison purposes, the PSNR values of the luminance component are measured and plotted over the total bit-rate for the quantization parameter 4, 5, 7, 10, 15, and 25. The data of the first intra-frame coded picture, which is identical in all cases, is excluded from the results.

4.3.1 Multiple Hypotheses for Constant Block Size

We investigate the coding efficiency of superimposed prediction with two and four hypotheses for constant block size. Figs. 4.1 and 4.2 depict the average luminance PSNR from reconstructed frames over the overall bit-rate for the sequences *Foreman* and *Mobile & Calendar*, respectively. The performance of the codec with baseline prediction (BL), superimposed prediction with two hypotheses (BL + INTER2H), and four hypotheses (BL + INTER2H + INTER4H) is shown. In each case, $M = 10$ reference pictures are utilized for prediction. The baseline performance for single frame prediction ($M = 1$) is added for reference.

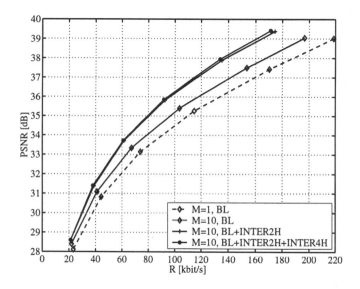

Figure 4.1. Average luminance PSNR over total rate for the sequence *Foreman* depicting the performance of the superposition coding scheme for constant block size. $M = 10$ reference pictures are utilized for prediction.

Superimposed prediction is enabled by allowing the INTER2H mode on the macroblock level. A gain of up to 1 dB for the sequence *Foreman* and 1.4 dB for the sequence *Mobile & Calendar* is achieved by the INTER2H mode. Superimposed prediction with up to four hypotheses is implemented such that each predictor type (depending on the number of superimposed signals) constitutes a coding mode. A rate-distortion efficient codec should utilize four hypotheses only when their coding gain is justified by the associated bit-rate. In the case that four hypotheses are not efficient, the codec should be able to select two hypotheses and choose the INTER2H mode. The additional INTER4H mode gains just up to 0.1 dB for the sequence *Foreman* and 0.3 dB for

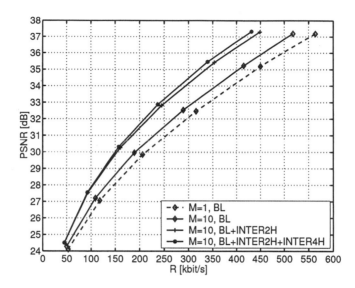

Figure 4.2. Average luminance PSNR over total rate for the sequence *Mobile & Calendar* depicting the performance of the superposition coding scheme for constant block size. $M = 10$ reference pictures are utilized for prediction.

the sequence *Mobile & Calendar*. This results support the finding in Section 3.2.2 that two hypotheses provide the largest relative gain. Considering both this insight and the computational complexity for estimating four hypotheses, we will restrict the superposition coding scheme to two hypotheses.

Finally, we consider hypotheses that are not optimized with the hypothesis selection algorithm in Fig. 2.6. With only previous reference frames, not optimized hypotheses are as good as or worse (due to the bit-rate of additional motion vectors) than single hypothesis prediction. In that case, the mode selection prefers the single hypothesis mode and the rate-distortion performance is identical to that with the label "$M = 10$, BL".

4.3.2   Multiple Hypotheses for Variable Block Size

In this subsection, we investigate the influence of variable block size (VBS) prediction on superimposed prediction for $M = 10$ reference pictures. VBS prediction in H.263 is enabled by the INTER4V mode which utilizes four motion vectors per macroblock. Both VBS prediction and superimposed prediction use more than one motion vector per macroblock which is transmitted to the decoder as side-information. But both concepts provide gains for different scenarios. This can be verified by applying superimposed prediction to blocks of size 16×16 (INTER2H) as well as 8×8 (INTER4VMH). As we permit a maximum of two hypotheses per block, one bit is sufficient to signal whether one or two prediction signals are used.

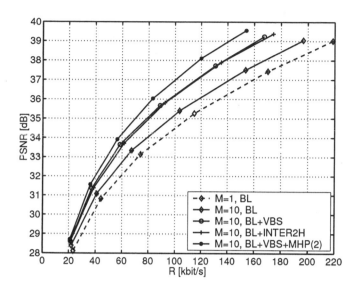

Figure 4.3. Average luminance PSNR over total rate for the sequence *Foreman*. Superimposed and variable block size prediction can be successfully combined for compression. $M = 10$ reference pictures are utilized for prediction.

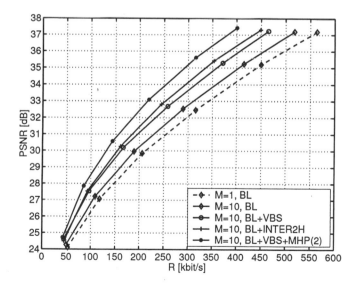

Figure 4.4. Average luminance PSNR over total rate for the sequence *Mobile & Calendar*. Superimposed and variable block size prediction can be successfully combined for compression. $M = 10$ reference pictures are utilized for prediction.

Figs. 4.3 and 4.4 depict the average luminance PSNR from reconstructed frames over the overall bit-rate for the sequences *Foreman* and *Mobile & Calendar*. The performance of the codec with baseline prediction (BL), VBS prediction (BL + VBS), superimposed prediction with two hypotheses (BL + INTER2H), and superimposed prediction with variable block size (BL + VBS + MHP(2)) is shown. In each case, $M = 10$ reference pictures are utilized for prediction. The baseline performance for single frame prediction ($M = 1$) is added for reference.

The combination of superimposed and variable block size prediction yields superior compression efficiency. For example, to achieve a reconstruction quality of 35 dB in PSNR, the sequence *Mobile & Calendar* is coded in baseline mode with 403 kbit/s for $M = 10$ (See Fig. 4.4). Correspondingly, superimposed prediction with $M = 10$ reduces the bit-rate to 334 kbit/s. Superimposed prediction on macroblocks decreases the bit-rate by 17%. Performing superimposed prediction additionally on 8×8 blocks, the bit-rate is 290 kbit/s in contrast to 358 kbit/s for the codec with VBS. Superimposed prediction decreases the bit-rate by 19% relative to our codec with VBS prediction. Similar observations can be made for the sequence *Foreman* at 120 kbit/s. Superimposed prediction on macroblocks gains about 1 dB over baseline prediction for $M = 10$ (See Fig. 4.3). Performing superimposed prediction additionally on 8×8 blocks, the gain is about 0.9 dB compared to the codec with VBS and $M = 10$ reference pictures.

Please note that the coding efficiency for the sequences *Foreman* (Fig. 4.3) and *Mobile & Calendar* (Fig. 4.4) is comparable for VBS prediction (BL + VBS) and superimposed prediction with two hypotheses (BL + INTER2H) over the range of bit-rates considered. Superimposed prediction utilizes just two motion vectors and picture reference parameters compared to four for the INTER4V mode.

For variable block size prediction, four hypotheses provide also no significant improvement over two hypotheses. For example, the superposition codec with VBS and four hypotheses achieves just up to 0.3 dB gain over the codec with two hypotheses for the sequence *Mobile & Calendar*.

In summary, superimposed prediction works efficiently for both 16×16 and 8×8 blocks. The savings due to superimposed prediction are observed in the baseline mode as well as in the VBS prediction mode. Hence, the hypothesis selection algorithm in Fig. 2.6 is able to find two prediction signals on $M = 10$ reference frames which are combined more efficiently than just one prediction signal from these reference frames.

4.3.3 Multiple Hypotheses and Multiple Reference Pictures

The results presented so far are obtained for superimposed motion-compensated prediction with $M = 10$ reference pictures in sliding window buffering mode. In this section, the influence of multiple reference frames on the superposition codec is investigated [210]. It is demonstrated that two hypotheses chosen only from the prior decoded frame, i.e. $M = 1$ also improve coding efficiency. Additionally, the use of multiple reference frames enhances the efficiency of the superposition codec.

Figs. 4.5 and 4.6 show the bit-rate savings at 35 dB PSNR of the decoded luminance signal over the number of reference frames M for the sequences *Foreman* and *Mobile & Calendar*, respectively. We compute PSNR vs. bit-rate curves by varying the quantization parameter and interpolate intermediate points by a cubic spline. The performance of the codec with variable block size prediction (VBS) is compared to the superposition codec with two hypotheses (VBS + MHP(2)). Results are depicted for $M = 1, 2, 5, 10$, and 20.

Figs. 4.7 and 4.8 show the same experiment as Figs. 4.5 and 4.6 but depict the absolute bit-rate over the number of reference pictures M for the sequences *Foreman* and *Mobile & Calendar*, respectively. The relative bit-rate savings with two hypotheses are given.

The superposition codec with $M = 1$ reference frame has to choose both prediction signals from the previous decoded frame. The superposition codec with VBS saves 7% for the sequence *Foreman* and 9% for the sequence *Mobile & Calendar* when compared to the VBS codec with one reference frame. For $M > 1$, more than one reference frame is allowed for each prediction signal. The reference frames for both hypotheses are selected by the rate-constrained superimposed motion estimation algorithm. The picture reference parameter allows also the special case that both hypotheses are chosen from the same reference frame. The rate constraint is responsible for the trade-off between prediction quality and bit-rate. For $M = 20$ reference frames, the superposition codec with VBS saves 25% for the sequence *Foreman* and 31% for the sequence *Mobile & Calendar* when compared to the VBS codec with one reference frame. For the same number of reference frames, the VBS codec saves about 15% for both sequences. The superposition codec with VBS benefits when being combined with multiframe prediction so that the savings are more than additive. The bit-rate savings saturate for 20 reference frames for both sequences.

In Section 3.4.4 we model multiframe motion compensation by forward-adaptive hypothesis switching. When being combined with complementary hypotheses motion compensation, we observe that the bit-rate savings by complementary hypotheses motion compensation over single hypothesis motion

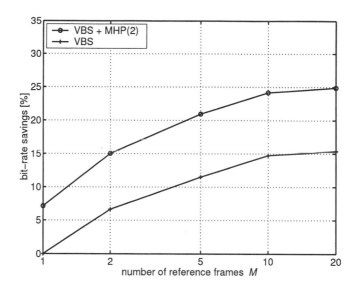

Figure 4.5. Bit-rate savings at 35 dB PSNR over the number of reference pictures M for the sequence *Foreman*. For variable block sizes, the performance of the superposition codec is compared to the reference codec.

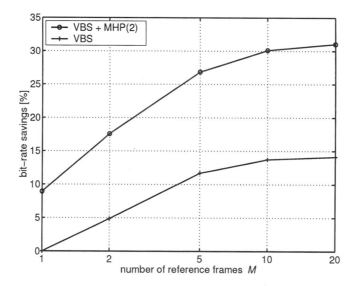

Figure 4.6. Bit-rate savings at 35 dB PSNR over the number of reference pictures M for the sequence *Mobile & Calendar*. For variable block sizes, the performance of the superposition codec is compared to the reference codec.

compensation increases with the number of motion-compensated signals M.

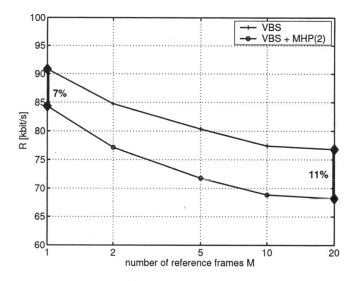

Figure 4.7. Absolute bit-rate at 35 dB PSNR over the number of reference pictures *M* for the sequence *Foreman*. For variable block sizes, the performance of the superposition codec is compared to the reference codec.

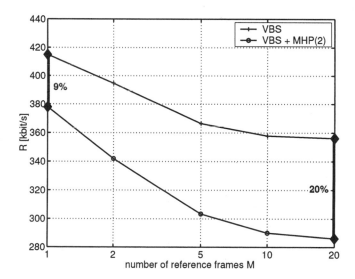

Figure 4.8. Absolute bit-rate at 35 dB PSNR over the number of reference pictures *M* for the sequence *Mobile & Calendar*. For variable block sizes, the performance of the superposition codec is compared to the reference codec.

For that, $N = 2$ complementary hypotheses are sufficient. This theoretical result is consistent with the previous experimental results.

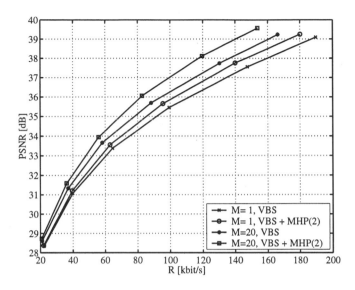

Figure 4.9.   Average luminance PSNR over total rate for the sequence *Foreman*. The performance of the superposition codec with variable block size is depicted for $M = 1$ and $M = 20$
reference frames.

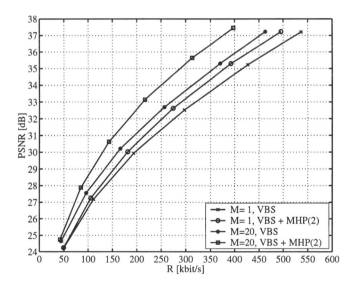

Figure 4.10.   Average luminance PSNR over total rate for the sequence *Mobile & Calendar*.
The performance of the superposition codec with variable block size is depicted for $M = 1$ and
$M = 20$ reference frames.

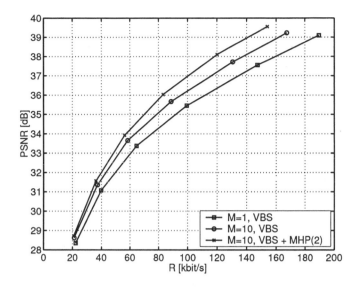

Figure 4.11. Average luminance PSNR over total rate for the sequence *Foreman*. The performance of the superposition codec with variable block size and multiframe motion compensation is compared to the reference codec.

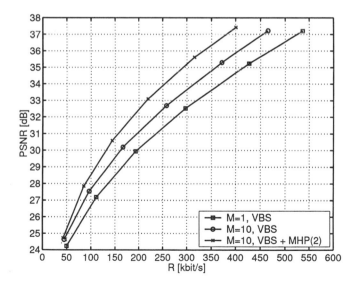

Figure 4.12. Average luminance PSNR over total rate for the sequence *Mobile & Calendar*. The performance of the superposition codec with variable block size and multiframe motion compensation is compared to the reference codec.

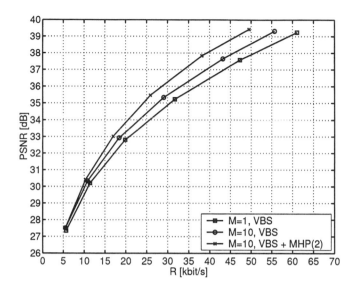

Figure 4.13.   Average luminance PSNR over total rate for the sequence *Sean*. The performance of the superposition codec with variable block size and multiframe motion compensation is compared to the reference codec.

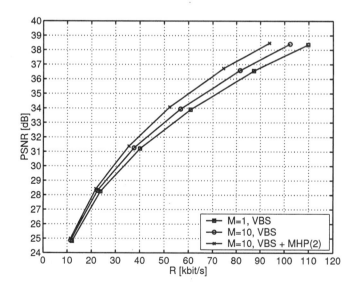

Figure 4.14.   Average luminance PSNR over total rate for the sequence *Weather*. The performance of the superposition codec with variable block size and multiframe motion compensation is compared to the reference codec.

Figs. 4.9 and 4.10 depict the average luminance PSNR over the total bit-rate for the sequences *Foreman* and *Mobile & Calendar*. The superposition codec with variable block size (VBS + MHP(2)) is compared to the variable block size codec (VBS) for $M = 1$ and $M = 20$ reference frames. We can observe in these figures that superimposed prediction in combination with multiframe motion compensation achieves coding gains up to 1.8 dB for *Foreman* and 2.8 dB for *Mobile & Calendar*. It is also observed that the use of multiple reference frames enhances the efficiency of superimposed motion-compensated prediction for video compression.

Finally, Figs. 4.5 and 4.6 suggest that a frame memory of $M = 10$ provides a good trade-off between encoder complexity and compression efficiency for our superposition codec. Therefore, Figs. 4.11, 4.12, 4.13, and 4.14 compare the superposition codec with variable block size and frame memory $M = 10$ to the reference codec with frame memory $M = 1$ and $M = 10$ for the sequences *Foreman*, *Mobile & Calendar*, *Sean*, and *Weather*, respectively. For each sequence the average luminance PSNR is depicted over the total bit-rate. The superposition codec with multiframe motion compensation achieves coding gains up to 1.8 dB for *Foreman*, 2.7 dB for *Mobile & Calendar*, 1.6 dB for *Sean*, and 1.5 dB for *Weather* compared to the reference codec with frame memory $M = 1$. The gain by multiframe prediction and superimposed prediction is comparable for the presented sequences.

4.4 Conclusions

In our experiments, we observe that variable block size and superimposed prediction can be combined successfully. Superimposed prediction works efficiently for both 16×16 and 8×8 blocks. Multiframe motion compensation enhances the efficiency of superimposed prediction. The superposition gain and the multiframe gain do not only add up; superimposed prediction benefits from hypotheses that can be chosen from different reference frames. Superimposed motion-compensated prediction with two hypotheses and ten reference frames achieves coding gains up to 2.7 dB, or equivalently, bit-rate savings up to 30% for the sequence *Mobile & Calendar* when compared to the reference codec with one reference frame. Therefore, superimposed prediction with multiframe and variable block size motion compensation is very efficient and practical for video compression.

Chapter 5

ITU-T REC. H.264 AND
GENERALIZED B-PICTURES

5.1   Introduction

This chapter discusses B-pictures in the context of the draft H.264 video compression standard. B-pictures are pictures in a motion video sequence that are encoded using both past and future pictures as references. The prediction is obtained by a linear combination of forward and backward prediction signals usually obtained with motion compensation. However, such a superposition is not necessarily limited to forward and backward prediction signals as investigated in Chapter 4. For example, a linear combination of two forward prediction signals can also be efficient in terms of compression efficiency. The prediction method which linearly combines motion-compensated signals regardless of the reference picture selection will be referred to as superimposed motion-compensated prediction. The concept of reference picture selection [91], also called multiple reference picture prediction, is utilized to allow prediction from both temporal directions. In this particular case, a bidirectional picture reference parameter addresses both past and future reference pictures [211]. This generalization in terms of picture reference selection and linearly combined prediction signals is reflected in the term *generalized B-pictures* and is part of the emerging H.264 video compression standard [212]. It is desirable that an arbitrary pair of reference pictures can be signaled to the decoder [213, 214]. This includes the classical combination of forward and backward prediction signals but also allows forward/forward as well as backward/backward pairs. When combining the two most previous pictures, a functionality similar to the dual-prime mode in MPEG-2 [13, 97] is achieved, where top and bottom fields are averaged to form the final prediction.

B-pictures in H.264 have been improved in several ways compared to B-pictures in MPEG-2 [13] and H.263 [14]. The block size for motion compensation can range from 16×16 to 4×4 pixels and the direct mode with weighted blending allows not only a scaling of the motion vectors but also a weighting of the prediction signal. The ongoing H.264 development will also provide improved H.263 Annex U functionality. H.263 Annex U, Enhanced Reference Picture Selection, already allows multiple reference pictures for forward prediction and two-picture backward prediction in B-pictures. When choosing between the most recent and the subsequent reference picture, the multiple reference picture selection capability is very limited. Utilizing multiple prior and subsequent reference pictures improves the compression efficiency of H.263 B-pictures.

The H.264 test model software TML-9 uses only inter pictures as reference pictures to predict the B-pictures. Beyond the test model software TML-9, and different from past standards, the multiple reference picture framework in H.264 also allows previously decoded B-pictures to be used as reference to improve prediction efficiency [215]. B-pictures can be utilized to establish an enhancement layer in a layered representation and allow temporal scalability [216]. That is, decoding of a sequence at more than one frame rate is achievable. In addition to this functionality, B-pictures generally improve the overall compression efficiency as compared to that of inter pictures only [217]. On the other hand, they increase the time delay due to multiple future reference pictures. But this disadvantage is not critical in applications like Internet streaming and multimedia storage for entertainment purposes.

The outline of this chapter is as follows: Section 5.2 introduces B-picture prediction modes. After an overview, direct and superposition mode are discussed in more detail and a rate-distortion performance comparison of three mode classes is provided. Section 5.3 elaborates on superimposed prediction. The difference between bidirectional and superposition mode is outlined and quantified in experimental results. In addition, the efficiency of two combined forward prediction signals is also investigated. Finally, both entropy coding schemes of H.264 are investigated with respect to the superposition mode. Encoder issues are detailed in Section 5.4, which covers rate-constrained mode decision, motion estimation, and superimposed motion estimation. In addition, the improvement of the overall rate-distortion performance with B-pictures is discussed.

5.2   Prediction Modes for B-Pictures

5.2.1   Overview

The macroblock modes for B-pictures allow intra and inter coding. The *intra-mode macroblocks* specified for inter pictures are also available for B-pictures. The *inter-mode macroblocks* are especially tailored to B-picture use. As for inter pictures, they utilize seven block size types as depicted in Fig. 5.1 to generate the motion-compensated macroblock prediction signal. In addition, the usage of the reference picture set available for predicting the current B-picture is suited to its temporally non-causal nature.

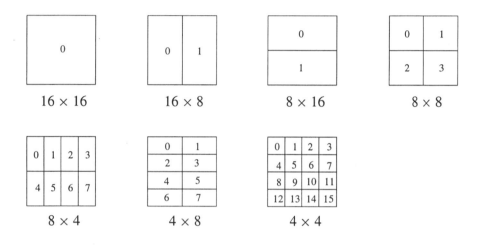

Figure 5.1.   Block size types for the motion-compensated macroblock prediction signal.

In contrast to the previously mentioned inter-mode macroblocks which signal motion vector data according to its block size as side information, the *direct-mode macroblock* does not require such side information but derives reference frame, block size, and motion vector data from the subsequent inter picture. This mode linearly combines two prediction signals. One prediction signal is derived from the subsequent inter picture, the other from a previous picture.

A linear combination of two motion-compensated prediction signals with explicit side information is accomplished by the *superposition mode*. Existing standards with B-pictures utilize the bidirectional mode, which only allows the combination of a previous and subsequent prediction signal. The superposition mode generalizes this concept and supports not only the already mentioned forward/backward prediction pair, but also forward/forward and backward/backward pairs.

5.2.2   Direct Mode

The direct mode uses bidirectional prediction and allows residual coding of the prediction error. The forward and backward motion vectors of this mode are derived from the motion vectors used in the corresponding macroblocks of the subsequent reference picture. The same number of motion vectors are used. To calculate prediction blocks, the forward and backward motion vectors are used to obtain appropriate blocks from reference pictures and then these blocks are linearly combined. Using multiple reference picture prediction, the forward reference picture for the direct mode is the same as the one used for the corresponding macroblock in the subsequent inter picture. The forward and backward motion vectors for direct mode macroblocks are calculated as

$$MV_F \;=\; \frac{TR_B}{TR_D} MV \qquad\qquad (5.1)$$

$$MV_B \;=\; \frac{TR_B - TR_D}{TR_D} MV, \qquad\qquad (5.2)$$

where MV_F is the forward motion vector, MV_B is the backward motion vector, and MV represents the motion vectors in the corresponding macroblock in the subsequent inter picture. TR_D is the temporal distance between the previous and the next inter picture, and TR_B is the distance between the current picture and the previous inter picture. It should be noted that when multiple reference picture prediction is used, the reference picture for the motion vector predictions is treated as though it were the most recent previous decoded picture. Thus, instead of using the temporal reference of the exact reference picture to compute the temporal distances TR_D and TR_B, the temporal reference of the most recent previous reference picture is used to compute the temporal distances.

The direct mode in H.264 is improved by weighted blending of the prediction signal [218]. Video content like music videos and movie trailers make frequent use of fading transitions from scene to scene. It is very popular in movie trailers to fade each scene to black, and then from black to the next scene. Without weighted blending of the prediction signal, both normal fades and "fades to-black" are hard to encode well without visible compression artifacts. For example, when encoding with a PBBB pattern, the B-pictures in position 1 and 3 suffer from quality degradation relative to the B-pictures in position 2 and the surrounding inter and intra pictures. The weighted blending technique considers how the direct mode motion vectors are derived from scaling the motion vector for the subsequent inter picture, based on the distance between the B-picture and the surrounding pictures, and also weighs the calculation of the prediction block based on this distance, instead of the averaging with equal weights that has been used in all existing standards with B-pictures. The weighted blending technique calculates the prediction block c for direct mode coded macroblocks according to

$$c = \frac{c_p(TR_D - TR_B) + c_s TR_B}{TR_D}, \tag{5.3}$$

where c_p is the prediction block from a previous reference picture, and c_s is the prediction block from the subsequent reference picture. Sequences without any fades will not suffer from loss of compression efficiency relative to the conventional way to calculate the prediction for direct mode.

5.2.3 Superposition Mode

The superposition mode superimposes two macroblock prediction signals with their individual sets of motion vectors. We refer to each prediction signal as a hypothesis. To calculate prediction blocks, the motion vectors of the two hypotheses are used to obtain appropriate blocks from reference pictures and then these blocks are averaged. Each hypothesis is specified by one of the seven block size types as depicted in Fig. 5.1. In addition, each hypothesis is also assigned one picture reference parameter. The motion vectors for each hypothesis are assigned on a block level and all of them refer to that specified reference picture. It is very likely that the hypotheses are chosen from different reference pictures but they can also originate from the same picture. Increasing the number of available reference pictures improves the performance of superimposed motion-compensated prediction as shown in Section 4.3.3 and theoretically discussed in Section 3.4.4. For B-pictures, more details are given in Section 5.3.

5.2.4 Rate-Distortion Performance of Individual Modes

The macroblock modes for B-pictures can be classified into four groups:

1. No extra side information is transmitted for this particular macroblock. This corresponds to the *direct* mode.

2. side information for one macroblock prediction signal is transmitted. The *inter* modes with block structures according to Fig. 5.1 and bidirectional picture reference parameters belong to this group.

3. side information for two macroblock prediction signals is transmitted to allow *superimposed* prediction.

4. The last group includes all *intra* modes, i.e., no inter-frame prediction is used.

In the following, the first three groups which utilize inter-frame prediction are investigated with respect to their rate-distortion performance. The fourth group with the intra modes is negligible. They are available in each experiment but their frequency is very small.

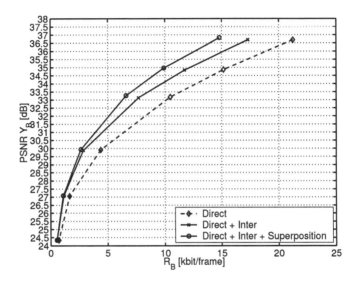

Figure 5.2. PSNR of the B-picture luminance signal vs. B-picture bit-rate for the QCIF sequence *Mobile & Calendar* with 30 fps. Two B-pictures are inserted after each inter picture. 5 past and 3 subsequent reference pictures are used. The compression efficiency of the B-picture coding modes *direct, inter,* and *superposition* are compared.

The rate-distortion performance of the groups *direct, inter*, and *superposition* are depicted in Fig. 5.2. The PSNR of the B-picture luminance signal is plotted over the B-picture bit-rate for the QCIF sequence *Mobile & Calendar*. With the direct mode for the B-pictures, the rate-distortion performance at high bit-rates is dominated by the efficiency of the residual encoding. The inter modes improve the compression efficiency approximately by 1 dB in PSNR at moderate and high bit-rates. At very low bit-rates, the rate-penalty in effect disables the modes in the inter group due to extra side information. Similar behavior can be observed for the superposition mode. Transmitting two prediction signals increases the side information. Consequently, the superposition mode improves compression efficiency approximately by 1 dB in PSNR at high bit-rates.

Corresponding to the rate-distortion performance of the three groups, Fig. 5.3 depicts the relative occurrence of the macroblock modes in B-pictures

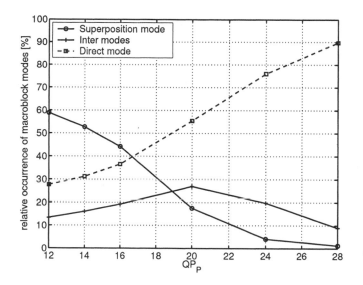

Figure 5.3. Relative occurrence of the macroblock modes in B-pictures vs. quantization parameter for the QCIF sequence *Mobile & Calendar* with 30 fps. Two B-pictures are inserted after each inter picture. 5 past and 3 subsequent reference pictures are used. The relative frequency of the B-picture macroblock modes *direct, inter,* and *superposition* are compared.

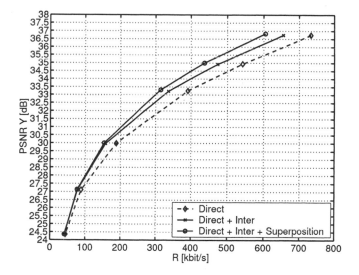

Figure 5.4. PSNR of the luminance signal vs. overall bit-rate for the QCIF sequence *Mobile & Calendar* with 30 fps. Two B-pictures are inserted after each inter picture. 5 past and 3 subsequent reference pictures are used. The compression efficiency of the B-picture coding modes *direct, inter,* and *superposition* are compared.

vs. quantization parameter QP_P for the QCIF sequence *Mobile & Calendar*. At $QP_P = 28$ (low bit-rate), the direct mode is dominant with approximately 90% relative occurrence, whereas the superposition and inter modes are seldom selected due to the rate-constraint. At $QP_P = 12$ (high bit-rate), the relative occurrence of the direct mode decreases to 30%, whereas the relative frequency of the superposition mode increases to 60%. About 10% of the macroblocks utilize an inter mode.

The influence of the B-picture coding modes *direct, inter,* and *superposition* on the overall compression efficiency is depicted in Fig. 5.4 for the QCIF sequence *Mobile & Calendar*. The base layer (the sequence of inter pictures) is identical in all three cases and only the B-picture coding modes are selected from the specified classes. For this sequence, the inter modes in the B-pictures improve the overall efficiency approximately by 0.5 dB. The superposition mode adds an additional 0.5 dB for higher bit-rates.

5.3   Superimposed Prediction

5.3.1   Bidirectional vs. Superposition Mode

In the following, we will outline the difference between the bidirectional macroblock mode, which is specified in the H.264 test model TML-9 [212], and the superposition mode proposed in [214] and discussed in the previous section. A bidirectional prediction type only allows a linear combination of a forward/backward prediction pair; see Fig. 5.5.

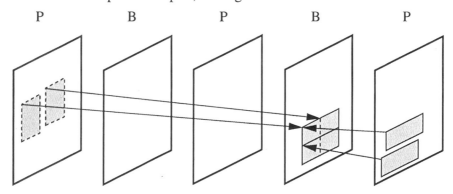

Figure 5.5.   A bidirectional prediction mode allows a linear combination of one past and one subsequent macroblock prediction signal. The inter pictures are denoted by P.

The draft TML-9 utilizes multiple reference pictures for forward prediction but allows only backward prediction from the most subsequent reference picture. For bidirectional prediction, independently estimated forward and backward prediction signals are practical but the efficiency can be improved by joint estimation. For superimposed prediction in general, a joint estimation of two hypotheses is necessary as discussed in Chapter 3. An independent estimate might even deteriorate the performance. The test model software TML-9 does not allow a joint estimation of forward and backward prediction signals.

The superposition mode removes the restriction of the bidirectional mode to allow only linear combinations of forward and backward pairs [219, 220]. The additional combinations (forward, forward) and (backward, backward) are obtained by extending an unidirectional picture reference syntax element to a bidirectional picture reference syntax element; see Fig. 5.6. With this bidirectional picture reference element, a generic prediction signal, which we call hypothesis, can be formed with the syntax fields for reference frame, block size, and motion vector data.

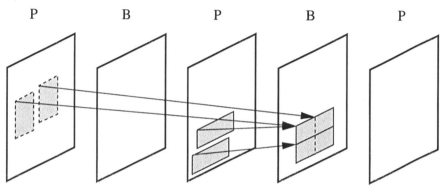

Figure 5.6. The superposition mode allows a linear combination of two past macroblock prediction signals (as depicted), two future macroblock signals, or one past and one future macroblock signal. The inter pictures are denoted by P.

The superposition mode includes the bidirectional prediction mode when the first hypothesis originates from a past reference picture and the second from a future reference picture. The bidirectional mode limits the set of possible reference picture pairs. Not surprisingly a larger set of reference picture pairs improves the coding efficiency of B-pictures.

The following results are based on the H.264 test model TML-9 [212]. For our experiments, the CIF sequences *Mobile & Calendar* and *Flowergarden* are coded at 30 fps. We investigate the rate-distortion performance of the superposition mode in comparison with the bidirectional mode when two B-pictures are inserted.

Figs. 5.7 and 5.8 depict the average luminance PSNR from reconstructed B-pictures over the overall bit-rate produced by B-pictures with bidirectional

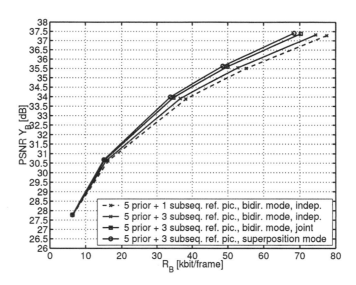

Figure 5.7. PSNR of the B-picture luminance signal vs. B-picture bit-rate for the CIF sequence *Mobile & Calendar* with 30 fps. Two B-pictures are inserted after each inter picture. $QP_B = QP_P$. The superposition mode is compared to the bidirectional mode.

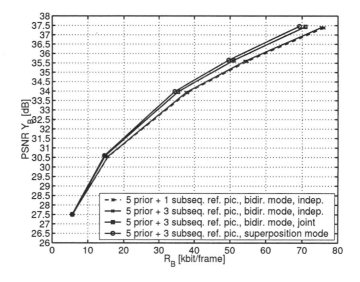

Figure 5.8. PSNR of the B-picture luminance signal vs. B-picture bit-rate for the CIF sequence *Flowergarden* with 30 fps. Two B-pictures are inserted after each inter picture. $QP_B = QP_P$. The superposition mode is compared to the bidirectional mode.

prediction mode and the superposition mode for the sequences *Mobile & Calendar* and *Flowergarden*. The number of reference pictures is chosen to be 1 and 3 future reference pictures with a constant number of 5 past pictures. It can be observed that increasing the total number of reference pictures from $5 + 1$ to $5 + 3$ slightly improves compression efficiency. Moreover, the superposition mode outperforms the bidirectional mode and its compression efficiency improves for increasing bit-rate. In the case of the bidirectional mode, jointly estimated forward and backward prediction signals outperform independently estimated signal pairs.

5.3.2   Two Combined Forward Prediction Signals

Generalized B-pictures combine both the superposition of prediction signals and the reference picture selection from past and future pictures. In the following, we investigate generalized B-pictures with forward-only prediction and utilize them like inter pictures in Chapter 4 for comparison purposes [213, 221]. That is, only a unidirectional reference picture parameter which addresses past pictures is permitted. As there is no future reference picture, the direct mode is replaced by the skip mode as specified for inter pictures. The *generalized B-pictures with forward-only prediction* cause no extra coding delay as they utilize only past pictures for prediction and are also used for reference to predict future pictures.

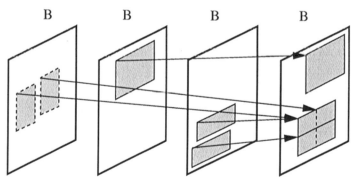

Figure 5.9.   Generalized B-pictures with forward-only prediction utilize multiple reference picture prediction and superimposed motion-compensated prediction. The superposition mode uses two hypotheses chosen from past reference pictures.

Fig. 5.9 shows generalized B-pictures with forward-only prediction. They allow multiple reference picture prediction and linearly combined motion-compensated prediction signals with individual block size types. Both hypotheses are just averaged to form the current macroblock. As depicted in Fig. 5.1, the H.264 test model [212] allows seven different block sizes which will be the seven hypotheses types in the superposition mode. The draft H.264

standard allows for inter modes only one picture reference parameter per macroblock and assumes that all sub-blocks can be found on that specified reference picture. This is different from the H.263 standard, where multiple reference picture prediction utilizes picture reference parameters for both macroblocks and 8 × 8 blocks [91].

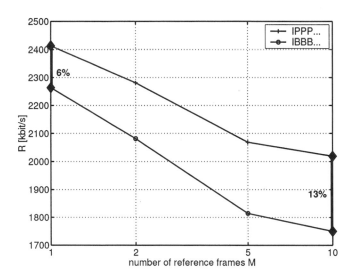

Figure 5.10.   Average bit-rate at 35 dB PSNR vs. number of reference pictures for the CIF sequence *Mobile & Calendar* with 30 fps. Generalized B-pictures with forward-only prediction are compared to inter pictures.

We investigate the rate-distortion performance of generalized B-pictures with forward-only prediction and compare them to H.264 inter pictures for various numbers of reference pictures. Figs. 5.10 and 5.11 show the bit-rate values at 35 dB PSNR of the luminance signal over the number of reference pictures M for the CIF sequences *Mobile & Calendar* and *Tempete*, respectively, coded at 30 fps. We compute PSNR vs. bit-rate curves by varying the quantization parameter and interpolate intermediate points by a cubic spline. The performance of H.264 inter pictures (IPPP...) and the generalized B-pictures with forward-only prediction (IBBB...) is shown.

The generalized B-pictures with forward-only prediction and $M = 1$ reference picture has to choose both hypotheses from the previous picture. For $M > 1$, we allow more than one reference picture for each hypothesis. The reference pictures for both hypotheses are selected by the rate-constrained superimposed motion estimation algorithm described in Section 5.4.3. The picture reference parameter allows also the special case that both hypotheses are chosen from the same reference picture. The rate constraint is responsible for

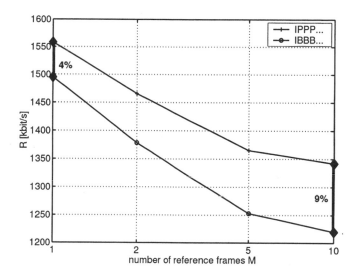

Figure 5.11. Average bit-rate at 35 dB PSNR vs. number of reference pictures for the CIF sequence *Tempete* with 30 fps. Generalized B-pictures with forward-only prediction are compared to inter pictures.

the trade-off between prediction quality and bit-rate. Using the generalized B-pictures with forward-only prediction and $M = 10$ reference pictures reduces the bit-rate from 2019 to 1750 kbit/s when coding the sequence *Mobile & Calendar*. This corresponds to 13% bit-rate savings. The gain by the generalized B-pictures with forward-only prediction and just one reference picture is limited to 6%. The gain by the generalized B-pictures over the inter pictures improves for a increasing number of reference pictures as already discussed in Section 4.3.3 for H.263. This observation is independent of the implemented superimposed prediction scheme and is supported by the theoretical investigation in Section 3.4.4.

Figs. 5.12 and 5.13 depict the average luminance PSNR from reconstructed pictures over the overall bit-rate produced by H.264 inter pictures (IPPP...) and the generalized B-pictures with forward prediction only (IBBB...) for the sequences *Mobile & Calendar* and *Tempete*, respectively. The number of reference pictures is chosen to be $M = 1$ and 5. It can be observed that the gain by generalized B-pictures improves for increasing bit-rate.

5.3.3   Entropy Coding

Entropy coding for H.264 B-pictures can be carried out in one of two different ways: universal variable length coding (UVLC) or context-based adap-

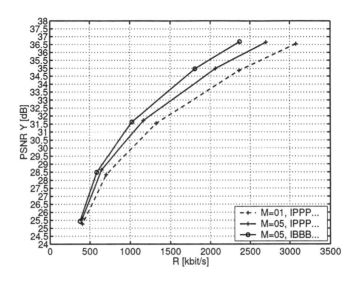

Figure 5.12. PSNR of the luminance signal vs. overall bit-rate for the CIF sequence *Mobile & Calendar* with 30 fps. Generalized B-pictures with forward-only prediction are compared to inter pictures.

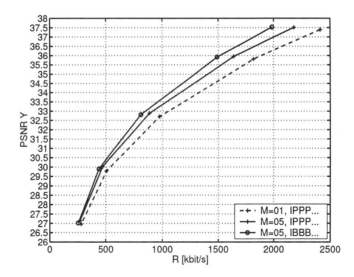

Figure 5.13. PSNR of the luminance signal vs. overall bit-rate for the CIF sequence *Tempete* with 30 fps. Generalized B-pictures with forward-only prediction are compared to inter pictures.

tive binary arithmetic coding (CABAC) [222–224]. The UVLC scheme uses only one variable length code to map all syntax elements to binary representations whereas CABAC utilizes context modeling and adaptive arithmetic codes to exploit conditional probabilities and non-stationary symbol statistics [223].

The simplicity of the UVLC scheme is striking as it demonstrates good compression efficiency at very low computational costs. CABAC with higher computational complexity provides additional bit-rate savings mainly for low and high bit-rates.

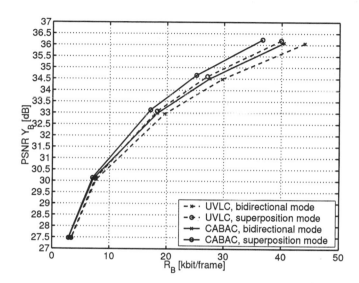

Figure 5.14. PSNR of the B-picture luminance signal vs. B-picture bit-rate for the CIF sequence *Mobile & Calendar* with 30 fps. Two B-pictures are inserted after each inter picture. 5 past and 3 future inter pictures are used for predicting each B-picture. $QP_B = QP_P + 2$ and $\lambda_B = 4w(QP_P)$. The superposition mode and the bidirectional mode with independent estimation are compared for both entropy coding schemes.

Figs. 5.14 and 5.15 depict the B-picture compression efficiency for the CIF sequences *Mobile & Calendar* and *Flowergarden*, respectively. For motion-compensated prediction, 5 past and 3 future inter pictures are used in all cases. The superposition mode and the bidirectional mode with independent estimation of prediction signals are compared for both entropy coding schemes. The PSNR gains by the superposition mode and the CABAC scheme are comparable for the investigated sequences at high bit-rates. When enabling the superposition mode together with CABAC, additive gains can be observed. The superposition mode improves the efficiency of motion-compensated prediction and CABAC optimizes the entropy coding of the utilized syntax elements.

The syntax elements used by the superposition mode can be coded with both the UVLC and the CABAC scheme. When using CABAC for the superposition mode, the context model for the syntax element motion vector data is adapted to superimposed motion. The context model for the motion vector of the first hypothesis captures the motion activity of the spatial neighbors, i.e. the left and

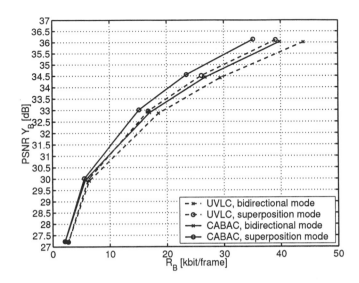

Figure 5.15. PSNR of the B-picture luminance signal vs. B-picture bit-rate for the CIF sequence *Flowergarden* with 30 fps. Two B-pictures are inserted after each inter picture. 5 past and 3 future inter pictures are used for predicting each B-picture. $QP_B = QP_P + 2$ and $\lambda_B = 4w(QP_P)$. The superposition mode and the bidirectional mode with independent estimation are compared for both entropy coding schemes.

top neighboring blocks. The context model for the motion vector of the second hypothesis captures the motion activity of the first hypothesis. The context models for the remaining syntax elements are not altered. The experimental results show that generalizing the bidirectional mode to the superposition mode improves B-picture compression efficiency for both schemes.

5.4   Encoder Issues

5.4.1   Rate-Constrained Mode Decision

The test model TML-9 distinguishes between a low- and high-complexity encoder. For a low-complexity encoder, computationally inexpensive rules for mode decision are recommended [225]. For a high-complexity encoder, the macroblock mode decision is ruled by minimizing the Lagrangian function

$$J_1(\text{Mode} \mid QP, \lambda) = SSD(\text{Mode} \mid QP) + \lambda R(\text{Mode} \mid QP), \qquad (5.4)$$

where QP is the macroblock quantizer parameter, and λ the Lagrange multiplier for mode decision. *Mode* indicates the selection from the set of potential coding modes. SSD is the sum of the squared differences between the original block and its reconstruction. It also takes into account the distortion in the chrominance components. R is the number of bits associated with choosing

Mode and QP, including the bits for macroblock header, motion information, and all integer transform blocks. The Lagrangian multiplier for λ is related to the macroblock quantizer parameter QP by

$$\lambda := w(QP) = 5\frac{QP+5}{34-QP} \exp\left(\frac{QP}{10}\right). \qquad (5.5)$$

Detailed discussions of this relationship can be found in [121] and [123]. Experimental results in Section 5.4.4 confirm that this relation should be adapted for B-pictures as specified in the test model TML-9,

$$\lambda_B = 4w(QP_B), \qquad (5.6)$$

such that the overall rate-distortion efficiency for the sequence is improved.

Mode decision selects the best mode among all B-picture macroblock modes. In order to reduce encoder complexity, mode decision assumes for all inter modes pre-computed motion vectors which are determined independently by rate-constrained motion estimation.

5.4.2   Rate-Constrained Motion Estimation

Motion estimation is also performed in a rate-constrained framework. The encoder minimizes the Lagrangian cost function

$$J_2(m, r \mid \lambda_{SAD}, p) = SAD(m, r) + \lambda_{SAD} R(m - p, r), \qquad (5.7)$$

with the motion vector m, the predicted motion vector p, the reference frame parameter r, and the Lagrange multiplier λ_{SAD} for the SAD distortion measure. The rate term R represents the motion information and the number of bits associated with choosing the reference picture r. The rate is estimated by table-lookup using the universal variable length code (UVLC) table, even if the arithmetic entropy coding method is used. For integer-pixel search, SAD is the summed absolute difference between the original luminance signal and the motion-compensated luminance signal. In the sub-pixel refinement search, the Hadamard transform of the difference between the original luminance signal and the motion-compensated luminance signal is calculated and SAD is the sum of the absolute transform coefficients. The Hadamard transform in the sub-pixel search reflects the performance of the integer transform on the residual signal such that the expected reconstruction quality rather than the motion-compensated prediction quality is taken into account for the refinement. This favors sub-pixel positions with residuals that are highly correlated for a given summed distortion. The Lagrangian multiplier λ_{SAD} for the SAD distortion measure is related to the Lagrangian multiplier for the SSD measure (5.5) by

$$\lambda_{SAD} = \sqrt{\lambda}. \qquad (5.8)$$

Further details as well as block size issues for motion estimation are discussed in [121] and [123].

5.4.3   Rate-Constrained Superimposed Motion Estimation

For the superposition mode, the encoder utilizes rate-constrained superimposed motion estimation. The cost function incorporates the superimposed prediction error of the video signal as well as the bit-rate for two picture reference parameters, two hypotheses types, and the associated motion vectors. Rate-constrained superimposed motion estimation is performed by the hypothesis selection algorithm in Fig. 2.6. This iterative algorithm performs conditional rate-constrained motion estimation and is a computationally feasible solution to the joint estimation problem [17] which has to be solved for finding an efficient pair of hypotheses.

The iterative algorithm is initialized with the data of the best macroblock type for multiple reference prediction (initial hypothesis). For two hypotheses, the algorithm continues with:

1. One hypothesis is fixed and conditional rate-constrained motion estimation is applied to the complementary hypothesis such that the superposition costs are minimized.

2. The complementary hypothesis is fixed and the first hypothesis is optimized.

The two steps (= one iteration) are repeated until convergence. For the current hypothesis, conditional rate-constrained motion estimation determines the conditional optimal picture reference parameter, hypothesis type, and associated motion vectors. For the conditional optimal motion vectors, an integer-pel accurate estimate is refined to sub-pel accuracy.

Fig. 5.16 shows the average number of iterations for superimposed motion estimation with 5 reference pictures over the quantization parameter. On average, it takes about 2 iterations to achieve a Lagrangian cost smaller than 0.5% relative to the cost in the previous iteration. The algorithm converges faster for higher quantization parameter values.

Given the best single hypothesis for motion-compensated prediction (best inter mode) and the best hypothesis pair for superimposed prediction, the resulting prediction errors are transform coded to compute the Lagrangian costs for the mode decision.

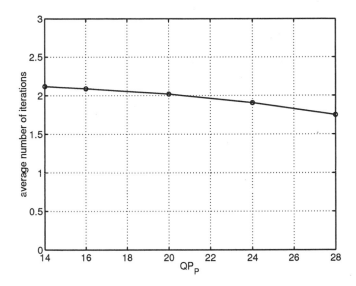

Figure 5.16. Average number of iterations for superimposed motion estimation vs. quantization parameter for the CIF sequence *Mobile & Calendar* with 30 fps and $M = 5$ reference pictures.

Superimposed motion-compensated prediction improves the prediction signal by allocating more bits to the side information associated with the motion-compensating predictor. But the encoding of the prediction error and its associated bit-rate also determines the quality of the reconstructed macroblock. A joint optimization of superimposed motion estimation and prediction error coding is far too demanding. But superimposed motion estimation independent of prediction error encoding is an efficient and practical solution if rate-constrained superimposed motion estimation is applied.

It turns out that the superposition mode is not necessarily the best one for each macroblock. Therefore, the rate-distortion optimal mode selection is a very important tool to decide whether a macroblock should be predicted with one or two hypotheses.

Fig. 5.17 shows the relative occurrence of the superposition mode in generalized B-pictures over the quantization parameter for the CIF sequence *Mobile & Calendar*. 5 past and 3 future reference pictures are used. Results for both entropy coding schemes are plotted. For high bit-rates (small quantization parameters), the superposition mode exceeds a relative occurrence of 50% among all B-picture coding modes. For low bit-rates (large quantization parameters), the superposition mode is selected infrequently and, consequently, the improvement in coding efficiency is very small. In addition, the relative occurrence is slightly larger for the CABAC entropy coding scheme since the

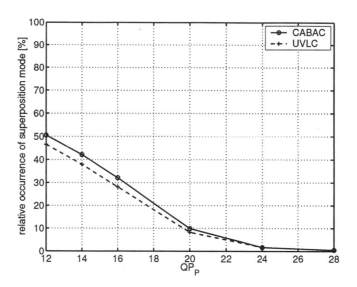

Figure 5.17. Relative occurrence of the superposition mode in B-pictures vs. quantization parameter for the CIF sequence *Mobile & Calendar* with 30 fps. 5 past and 3 future reference pictures are used. $QP_B = QP_P + 2$.

more efficient CABAC scheme somewhat relieves the rate constraint imposed on the side information.

5.4.4   Improving Overall Rate-Distortion Performance

When B-pictures are considered to be an enhancement layer in a scalable representation, they are predicted from reference pictures that are provided by the base layer. Consequently, the quality of the base layer influences the rate-distortion trade-off for B-pictures in the enhancement layer. Experimental results show that the relationship between quantization and Lagrange parameter for mode decision $\lambda = w(QP)$ should be adapted [226]. The following experimental results are obtained with the test model software TML-9, i.e., with bidirectional prediction and independent estimation of forward and backward prediction parameters.

Fig. 5.18 shows the PSNR of the luminance signal vs. overall bit-rate for the QCIF sequence *Mobile & Calendar* with 30 fps. Three different $\lambda - QP$ dependencies are depicted. The worst compression efficiency is obtained with $\lambda_B = w(QP_B)$. The cases $\lambda_B = 4w(QP_B)$ and $\lambda_B = 8w(QP_B)$ demonstrate superior efficiency for low bit-rates. The scaling of the dependency alters the bit-rate penalty for all B-picture coding modes such that the overall compres-

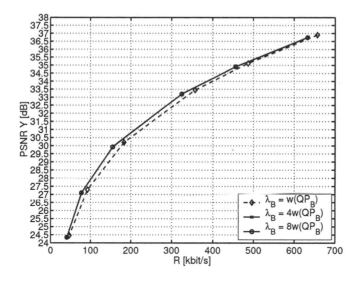

Figure 5.18. PSNR of the luminance signal vs. overall bit-rate for the QCIF sequence *Mobile & Calendar* with 30 fps. Two B-pictures are inserted and the influence of the $\lambda_B - QP_B$ relationship on the overall compression efficiency is investigated.

sion efficiency is improved. The factor 4 is suggested in the current test model description.

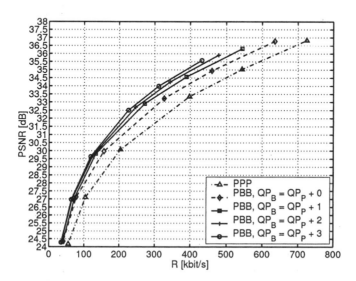

Figure 5.19. PSNR of the luminance signal vs. overall bit-rate for the QCIF sequence *Mobile & Calendar* with 30 fps. Two B-pictures are inserted and the influence of the B-picture quantization parameter QP_B on the overall compression efficiency is investigated for $\lambda_B = 4w(QP_B)$.

Further experiments show that not only the relationship between quantization and Lagrange parameter for mode decision has to be adapted for B-pictures but also the PSNR of the enhancement layer should be lowered in comparison to the base layer to improve overall compression efficiency [226]. Fig. 5.19 depicts also the PSNR of the luminance signal vs. overall bit-rate for the QCIF sequence *Mobile & Calendar* with 30 fps. The plot compares the compression efficiency of various layered bit-streams with two inserted B-pictures. The quantization parameters of inter and B-pictures differ by a constant offset. For comparison, the efficiency of the single layer bit-stream is provided. Increasing the quantization parameter for B-pictures, that is, lowering their relative PSNR, improves the overall compression efficiency of the sequence.

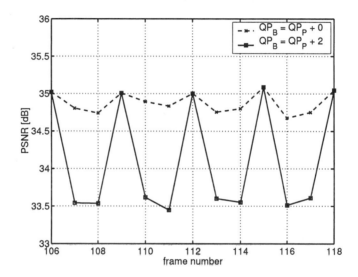

Figure 5.20. PSNR of the luminance signal for individual pictures. Two B-pictures are inserted. The B-picture quantization parameter QP_B is incremented by 2 and the B-picture Lagrange parameter $\lambda_B = 4w(QP_B)$. $QP_P = 14$.

Fig. 5.20 shows the PSNR of the luminance signal for individual pictures of the sequence *Mobile & Calendar* encoded with $QP_P = 14$. The PSNR of the B-pictures encoded with an increment of 2 is significantly lower compared to the case with identical quantization parameter in both layers. The compression efficiency of the sequence increases by lowering the relative PSNR of the enhancement layer. For the investigated sequence, the average PSNR efficiency increases by almost 1 dB (see Fig. 5.19), whereas the PSNR of individual B-pictures drops by more than 1 dB. In this case, higher average PSNR with temporal fluctuations is compared to lower average PSNR with less fluctuations for a given bit-rate.

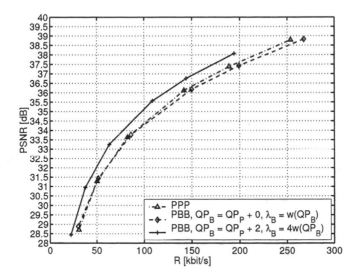

Figure 5.21. PSNR of the luminance signal vs. overall bit-rate for the QCIF sequence *Foreman* with 30 fps. When replacing two inter pictures by B-pictures, the values QP_B and λ_B have to be adapted for best compression efficiency. Keeping the inter picture values may lower the efficiency.

Fig. 5.21 shows the PSNR of the luminance signal vs. overall bit-rate for the QCIF sequence *Foreman* with 30 fps. The depicted results demonstrate that without adapting the quantization parameter and the $\lambda - QP$ dependency for B-pictures, we observe a degradation in compression efficiency if two inter pictures are replaced by B-pictures, whereas the adaptation improves the PSNR by about 0.5 dB for a given bit-rate.

5.5 Conclusions

This chapter discusses B-pictures in the context of the draft H.264 video compression standard. We focus on reference picture selection and linearly combined motion-compensated prediction signals. We show that bidirectional prediction only partially exploits the efficiency of combined prediction signals whereas superimposed prediction allows a more general form of B-pictures. The general concept of linearly combined prediction signals chosen from an arbitrary set of reference pictures further improves the H.264 test model TML-9 which is used in this chapter.

We outline H.264 macroblock prediction modes for B-pictures, classify them into four groups and compare their efficiency in terms of rate-distortion performance. When investigating superimposed prediction, we show that bidirectional prediction is a special case of this concept. Superimposed prediction allows also two combined forward prediction signals. Experimental results show that this case is also advantageous in terms of compression efficiency. The draft H.264 video compression standard offers improved entropy coding by context-based adaptive binary arithmetic coding. Simulations show that the gains by superimposed prediction and arithmetic coding are additive. B-pictures establish an enhancement layer and are predicted from reference pictures that are provided by the base layer. The quality of the base layer influences the rate-distortion trade-off for B-pictures. We demonstrate how the quality of the B-pictures should be reduced in order to improve the overall rate-distortion performance of the scalable representation.

Conceptually, we differentiate between picture reference selection and linearly combined prediction signals. This distinction is reflected in the term *Generalized B-Pictures*. The feature of reference picture selection has been improved significantly when compared to existing video compression standards. But the draft H.264 video compression standard has been extended recently to support the features of linearly combined prediction signals that are described in this chapter.

Towards a generalized picture type, a desirable definition would make a distinction whether only past reference pictures or also future reference pictures are used for prediction. If both past and future reference pictures are available, this generalized picture type would utilize the direct and superposition mode, whereas if only past reference frames are available, this generalized picture type would replace the direct mode by the copy mode.

Chapter 6

MOTION COMPENSATION
FOR GROUPS OF PICTURES

6.1 Introduction

So far, we discussed video coding schemes that utilize inter-frame methods with motion-compensated prediction (MCP) for efficient compression. Such compression schemes require sequential processing of video signals which makes it difficult to achieve efficient embedded representations of video sequences. This chapter investigates the efficiency of motion-compensated 3D transform coding, a compression scheme that employs a motion-compensated transform for groups of pictures. We investigate this coding scheme experimentally and theoretically. The practical coding scheme employs in temporal direction a motion-compensated subband decomposition for each group of pictures. We also compare the experimental results to that of a predictive video codec with motion compensation and comparable complexity. The theoretical investigation models this motion-compensated subband coding scheme for a group of K pictures with a signal model for K motion-compensated pictures that are decorrelated by a linear transform. We utilize the Karhunen-Loeve Transform to obtain theoretical performance bounds at high bit-rates and compare to both optimum intra-frame coding of individual motion-compensated pictures and motion-compensated predictive coding.

The outline of this chapter is as follows: Section 6.2 discusses the motion-compensated subband coding scheme with a motion-compensated lifted Haar wavelet and a motion-compensated lifted 5/3 wavelet. Experimental results and comparisons for several test sequences are presented. The theoretical signal model is developed in Section 6.3. The motion-compensated wavelet kernels of the practical coding schemes are generalized to obtain theoretical performance bounds.

6.2 Coding Scheme

Applying a linear transform in temporal direction of a video sequence is not very efficient if significant motion is prevalent. However, a combination of a linear transform and motion compensation seems promising for efficient compression. For wavelet transforms, the so called *Lifting Scheme* [156] can be used to construct the kernels. A two-channel decomposition can be achieved with a sequence of prediction and update steps that form a ladder structure. The advantage is that this lifting structure is able to map integers to integers without requiring invertible lifting steps. Further, motion compensation can be incorporated into the prediction and update steps as proposed in [20]. The fact that the lifting structure is invertible without requiring invertible lifting steps makes this approach feasible. We cannot count on motion compensation to be invertible in general. If it is invertible, this motion-compensated wavelet transform based on lifting permits a linear transform along the motion trajectories in a video sequence.

In the following, we investigate coding schemes that process video sequences in groups of K pictures (GOP). First, we decompose each GOP in temporal direction. The dyadic decomposition utilizes a motion-compensated wavelet which will be discussed later in more detail. The temporal transform provides K output pictures that are intra-frame encoded. In order to allow a comparison to a basic predictive coder with motion compensation, we utilize for the intra-frame coder a 8×8 DCT with run-length coding. If we employ a Haar wavelet and set the motion vectors to zero, the dyadic decomposition will be an orthonormal transform. Therefore, we select the same quantizer stepsize for all K intra-frame encoder. The motion information that is required for the motion-compensated wavelet transform is estimated in each decomposition level depending on the results of the lower level. Further, we employ half-pel accurate motion compensation with bi-linear interpolation.

6.2.1 Motion-Compensated Lifted Haar Wavelet

First, we discuss the lifting scheme with motion compensation for the Haar wavelet [20]. Fig. 6.1 depicts a Haar transform with motion-compensated lifting steps. The even frames of the video sequence $s_{2\kappa}$ are used to predict the odd frames $s_{2\kappa+1}$ (*prediction step P*). The prediction step is followed by an *update step U*.

If the motion field between the even and odd frames is invertible, the corresponding motion vectors in the update and prediction steps sum to zero. We use a block-size of 16×16 and half-pel accurate motion compensation in the prediction step and select the motion vectors such that they minimize the squared

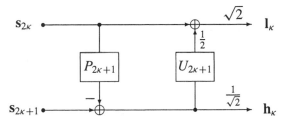

Figure 6.1. Haar transform with motion-compensated lifting steps. Both steps, prediction P and update U, utilize block-based motion compensation.

error in the high-band $\mathbf{h}_\kappa$. In general, this block-motion field is not invertible but we still utilize the negative motion vectors for the update step as an approximation. Additional scaling factors in low- and high-band are necessary to normalize the transform.

6.2.2 Motion-Compensated Lifted 5/3 Wavelet

The Haar wavelet is a short filter and provides limited coding gain. We expect better coding efficiency with longer wavelet kernels. In the following, we discuss the lifted 5/3 wavelet kernel with motion compensation [20].

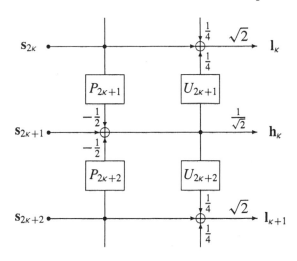

Figure 6.2. Lifted 5/3 wavelet with motion compensation.

Fig. 6.2 depicts the 5/3 transform with motion-compensated lifting steps. Similar to the Haar transform, the update steps $U_{2\kappa+1}$ use the negative motion vectors of the corresponding prediction steps. But for this transform, the odd frames are predicted by a linear combination of two displaced neighboring even frames. Again, we use a block-size of 16×16 and half-pel accurate motion compensation in the prediction steps and choose the motion vectors for $P_{2\kappa+1}$ and $P_{2\kappa+2}$ such that they minimize the squared error in the high-band $\mathbf{h}_\kappa$. The corresponding update steps $U_{2\kappa+2}$ use also the negative motion vectors of the corresponding prediction steps.

6.2.3 Experimental Results

For the experiments, we subdivide the QCIF sequences *Mother & Daughter*, *Container Ship*, *Salesman*, *Mobile & Calendar*, *Foreman*, *News*, and *Car Phone*, each with 288 frames at 30 fps, into groups of K pictures. We decompose the GOPs independently and in the case of the 5/3 wavelet, we refer back to the first picture in the GOP when the GOP terminates. We will justify later, why we choose this cyclic extension to handle the GOP boundaries.

Figs. 6.3, 6.4, 6.5, 6.6, 6.7, 6.8, and 6.9 show luminance PSNR over the total bit-rate for the test sequences encoded for groups of $K = 2, 8, 16$, and 32 pictures with the Haar kernel and for groups of $K = 32$ with the 5/3 kernel.

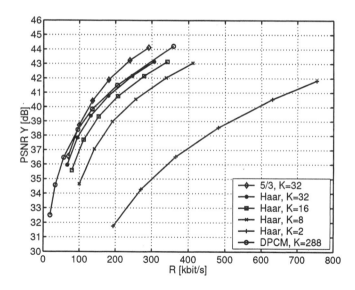

Figure 6.3. Luminance PSNR vs. total bit-rate for the QCIF sequence *Mother & Daughter* at 30 fps. A dyadic decomposition is used to encode groups of $K = 2, 8, 16$, and 32 pictures with the Haar kernel, and $K = 32$ with the 5/3 kernel. Results for a basic predictive video codec with 287 inter-frames are given for reference.

We observe that the bit-rate savings with the Haar kernel diminish very quickly as the GOP size approaches 32 pictures. Note also that the 5/3 decomposition with a GOP size of 32 outperforms the Haar decomposition with a GOP size of 32. For the sequences *Mother & Daughter, Container Ship,*

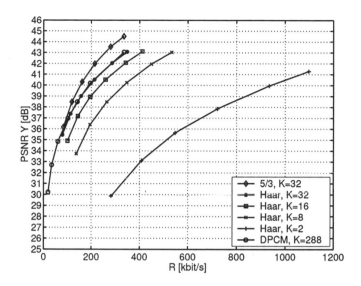

Figure 6.4. Luminance PSNR vs. total bit-rate for the QCIF sequence *Container Ship* at 30 fps. A dyadic decomposition is used to encode groups of $K = 2, 8, 16$, and 32 pictures with the Haar kernel, and $K = 32$ with the 5/3 kernel. Results for a basic predictive video codec with 287 inter-frames are given for reference.

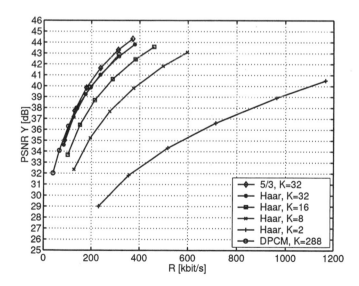

Figure 6.5. Luminance PSNR vs. total bit-rate for the QCIF sequence *Salesman* at 30 fps. A dyadic decomposition is used to encode groups of $K = 2, 8, 16$, and 32 pictures with the Haar kernel, and $K = 32$ with the 5/3 kernel. Results for a basic predictive video codec with 287 inter-frames are given for reference.

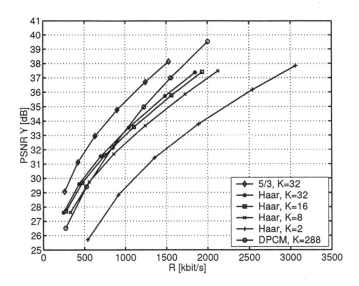

Figure 6.6.   Luminance PSNR vs. total bit-rate for the QCIF sequence *Mobile & Calendar* at 30 fps. A dyadic decomposition is used to encode groups of $K = 2, 8, 16$, and 32 pictures with the Haar kernel, and $K = 32$ with the 5/3 kernel. Results for a basic predictive video codec with 287 inter-frames are given for reference.

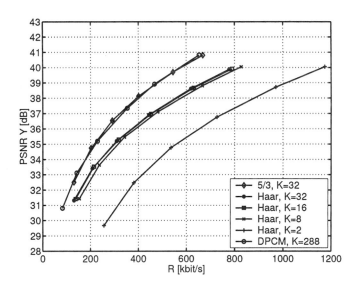

Figure 6.7.   Luminance PSNR vs. total bit-rate for the QCIF sequence *Foreman* at 30 fps. A dyadic decomposition is used to encode groups of $K = 2, 8, 16$, and 32 pictures with the Haar kernel, and $K = 32$ with the 5/3 kernel. Results for a basic predictive video codec with 287 inter-frames are given for reference.

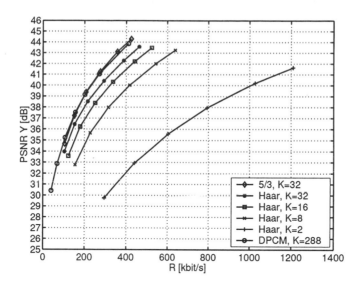

Figure 6.8. Luminance PSNR vs. total bit-rate for the QCIF sequence *News* at 30 fps. A dyadic decomposition is used to encode groups of $K = 2, 8, 16$, and 32 pictures with the Haar kernel, and $K = 32$ with the 5/3 kernel. Results for a basic predictive video codec with 287 inter-frames are given for reference.

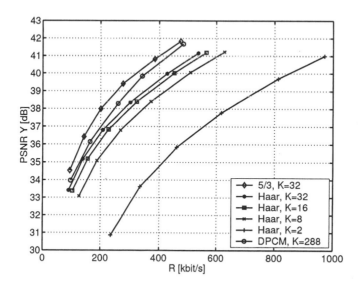

Figure 6.9. Luminance PSNR vs. total bit-rate for the QCIF sequence *Car Phone* at 30 fps. A dyadic decomposition is used to encode groups of $K = 2, 8, 16$, and 32 pictures with the Haar kernel, and $K = 32$ with the 5/3 kernel. Results for a basic predictive video codec with 287 inter-frames are given for reference.

and *Salesman*, the Haar wavelet coding scheme with $K = 32$ performs similar to a comparable basic predictive video codec (intra- and inter-frames, 16×16 block-motion compensation, half-pel accuracy, previous reference picture only, and 8×8 DCT) with a very large GOP size. Please note that for *Mobile & Calendar* at lower bit-rates the Haar wavelet coding scheme outperforms the predictive video codec. The 5/3 decomposition with a GOP size of 32 outperforms not only the corresponding Haar decomposition but also the basic predictive video coding scheme with a GOP size of $K = 288$. For the sequences *Foreman*, *News*, and *Car Phone*, the 5/3 wavelet coding scheme performs comparable or slightly better than the predictive video codec. These sequences contain inhomogeneous motion and we suspect that the use of negative motion vectors in the update step permits only an insufficient approximation.

Further, we investigate the behavior of the coding scheme for the cases that the frames are degraded by additive noise. For that, we generate the sequence *Noisy Foreman* by repeating 32 times the first frame of the sequence *Foreman* and adding statistically independent white Gaussian noise of variance 25. As we investigate the residual noise only, this sequence contains no motion. Predictive coding with motion compensation is not capable of predicting the additive noise in the current frame. In fact, prediction doubles the noise variance in the residual signal and we expect that predictive coding performs inferior.

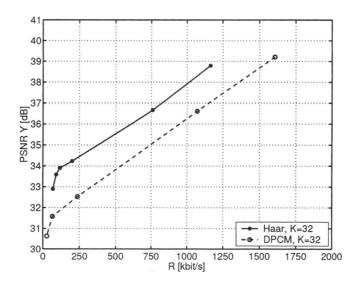

Figure 6.10. Luminance PSNR vs. total bit-rate for the QCIF sequence *Noisy Foreman* at 30 fps. A dyadic decomposition is used to encode groups of $K = 32$ pictures with the motion-compensated Haar wavelet. Results for a basic predictive video codec with 31 inter-frames are given for reference.

Fig. 6.10 shows luminance PSNR over the total bit-rate for the sequence *Noisy Foreman*. The coding scheme with the Haar wavelet kernel and a dyadic decomposition of 32 pictures is compared to the predictive coding scheme. We observe that the wavelet coding scheme outperforms the predictive coding scheme by approximately 2 dB. The predictive coding scheme is inferior as the statistically independent noise in the current frame cannot be predicted from the previous frame.

Finally, we discuss briefly our GOP boundary handling for the 5/3 wavelet. As we encode the GOPs in the sequence independently, we have to solve the boundary problem for the 5/3 wavelet. For the discussion, we consider cyclic and symmetric extensions at the GOP boundary. Note that a GOP begins with the even picture s_0 and terminates always with an odd picture. When the GOP terminates, the cyclic extension refers back to the first picture s_0, and the symmetric extension uses the last even picture in the GOP twice as a reference. In the case of the cyclic extension, the terminal update step modifies the first picture s_0, and in the case of the symmetric extension, the last even picture in the GOP is updated twice. We implemented both extensions for an experimental comparison of the rate-distortion performance.

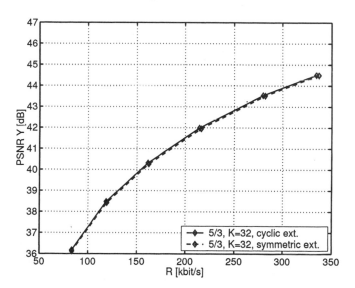

Figure 6.11. Luminance PSNR vs. total bit-rate for the QCIF sequence *Container Ship* at 30 fps. A dyadic decomposition is used to encode groups of $K = 32$ pictures with the motion-compensated 5/3 wavelet. The compression efficiency of cyclic and symmetric extension is depicted.

Fig. 6.11 shows luminance PSNR over the total bit-rate for the sequence *Container Ship*. We observe that both extensions show similar rate-distortion performance. We expect better efficiency for the cyclic extension as it is a multi-frame approach. But it seems that the large temporal distance is disadvantageous. As we observe a small advantage in terms of compression efficiency, also for other sequences like *Foreman*, we use the cyclic extension for our experiments.

6.3   Mathematical Model of Motion-Compensated Subband Coding

The experimental results show that the temporal subband coding scheme can provide superior compression efficiency when compared to the predictive coding scheme. In the following, we outline a mathematical model to study motion-compensated subband coding in more detail. With that, we derive performance bounds for motion-compensated three-dimensional transform coding and compare to bounds known for motion-compensated predictive coding.

Let $\mathbf{s}_k = \{\mathbf{s}_k[l], l \in \Pi\}$ be scalar random fields over a two-dimensional orthogonal grid Π with horizontal and vertical spacing of 1. The vector $l = (x, y)^T$ denotes a particular location in the lattice Π. We interpret $\mathbf{s}_k$ as the k-th of K pictures to be encoded. Further, the signal $\mathbf{s}_k[l]$ is thought of as samples of a space-continuous, spatially band-limited signal and we obtain a displaced version of it as follows: We shift the ideal reconstruction of the band-limited signal by the continuous-valued displacement vector d and re-sample it on the original grid. With this signal model, a spatially constant displacement operation is invertible.

6.3.1   Motion-Compensated Lifted Haar Wavelet

With the above signal model, we revisit the motion-compensated lifted Haar wavelet in Fig. 6.1 and remove the displacement operators in the lifting steps such that we can isolate a lifted Haar wavelet without displacement operators.

Fig. 6.12 shows the equivalent Haar wavelet where the displacement operators are pre- and post-processing operators with respect to the original Haar transform. The schemes in Fig. 6.12 are equivalent, if the displacement operators are linear and invertible.

We continue and perform the dyadic decomposition of a GOP with the equivalent Haar wavelet. For that, the displacements of the equivalent Haar blocks have to be added. We assume that the estimated displacements between pairs of frames are additive such that, e.g., $\hat{d}_{02} + \hat{d}_{23} = \hat{d}_{03}$. As the true displacements are also additive, e.g. $d_{02} + d_{23} = d_{03}$, and differ from the estimated displacement by the displacement error, i.e. $d_{ij} = \hat{d}_{ij} + \Delta_{ij}$, we conclude that the displacement errors are also additive, e.g. $\Delta_{02} + \Delta_{23} = \Delta_{03}$, [227].

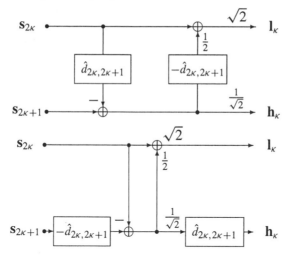

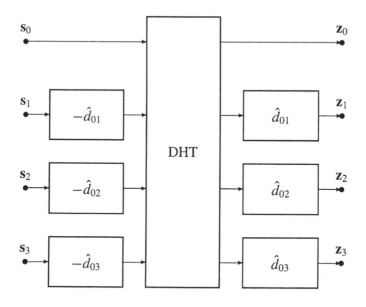

Figure 6.12. Haar transform with lifting steps that shift the signal (top). As the shift operation is invertible, an equivalent system without shifts in the lifting steps is possible (bottom).

Figure 6.13. Dyadic Haar Transform (DHT) without shifts in the lifting steps for $K = 4$ pictures.

Fig. 6.13 depicts a dyadic decomposition for $K = 4$ pictures based on the equivalent Haar wavelet in Fig. 6.12. The dyadic Haar transform without displacements in the lifting steps is labeled by DHT. The displacements $\hat{d}_{0k}$ are pre- and post-processing operators with respect to the original dyadic Haar decomposition DHT.

6.3.2   Motion-Compensated Lifted 5/3 Wavelet

We also apply the invertible displacement operator to the motion-compensated lifted 5/3 wavelet in Fig. 6.2 and obtain the equivalent 5/3 wavelet in Fig. 6.14.

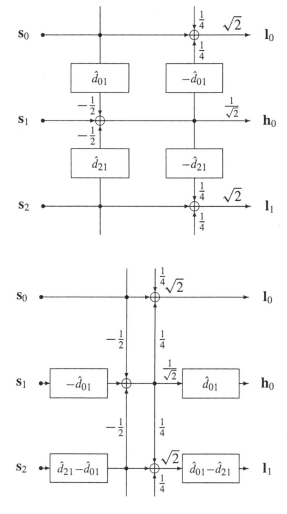

Figure 6.14.   5/3 wavelet with lifting steps that shift the signal (top). As the shift operation is invertible, an equivalent system without shifts in the lifting steps is possible (bottom).

Due to the structure of the 5/3 wavelet, we have displacements between the frames 2κ & $2\kappa + 1$, $2\kappa + 2$ & $2\kappa + 1$, and 2κ & $2\kappa + 2$ (in the next decomposition level). Again, we assume that the estimated displacements are additive such that, e.g., $\hat{d}_{01} - \hat{d}_{21} = \hat{d}_{02}$. With this assumption, the displacement operators between the levels cancel out and several decomposition levels are possible without displacements between the levels.

The equivalent dyadic 5/3 transform has the same pre- and post-processing displacement operators as the equivalent dyadic Haar transform in Fig. 6.13 but the DHT is replaced by the original dyadic 5/3 decomposition as depicted in Fig. 6.15.

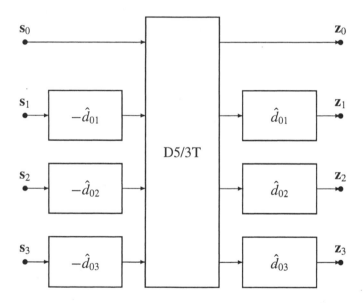

Figure 6.15.   Dyadic 5/3 Transform (D5/3T) without shifts in the lifting steps for $K = 4$ pictures.

6.3.3   Signal Model

Now, we assume that the pictures $\mathbf{s}_k$ are shifted versions of a "clean" video signal $\mathbf{v}$ with the true displacements d_{0k} and distorted by independent additive white Gaussian noise $\mathbf{n}_k$. Combining this signal model with the equivalent dyadic decomposition, we can eliminate the absolute displacements and restrict ourselves to the displacement error Δ_{0k} in the k-th picture. In the following, we do not consider particular displacement errors Δ_{0k}. We rather specify statistical properties and consider them as random variables $\mathbf{\Delta}_k$, statistically independent from the "clean" signal $\mathbf{v}$ and the noise $\mathbf{n}_k$. The noise signals $\mathbf{n}_\mu$ and $\mathbf{n}_\nu$ are also mutually statistically independent.

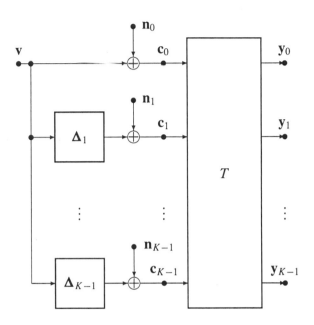

Figure 6.16.   Motion compensation for a group of K pictures.

Fig. 6.16 depicts the generalized model with the displacement-free and linear transform T for a group of K pictures. The motion-compensated pictures $\mathbf{c}_1, \ldots, \mathbf{c}_{K-1}$ are aligned with respect to the first picture $\mathbf{c}_0$. According to Fig. 6.13, the signals $\mathbf{z}_k$ are independently intra-frame encoded. As the absolute displacements have no influence on the performance of the intra-frame encoder, we omit them and consider only the direct output signals $\mathbf{y}_k$ of T.

Now, assume that the random fields $\mathbf{v}$ and $\mathbf{c}_k$ are jointly wide-sense stationary with the real-valued scalar two-dimensional power spectral densities $\Phi_{\mathbf{vv}}(\omega)$ and $\Phi_{\mathbf{c}_\mu \mathbf{c}_\nu}(\omega)$. The power spectral densities $\Phi_{\mathbf{c}_\mu \mathbf{c}_\nu}(\omega)$ are elements in the power spectral density matrix of the motion-compensated pictures $\Phi_{\mathbf{cc}}$. The power spectral density matrix of the decorrelated signal $\Phi_{\mathbf{yy}}$ is given by $\Phi_{\mathbf{cc}}$ and the transform T,

$$\Phi_{\mathbf{yy}}(\omega) = T(\omega)\Phi_{\mathbf{cc}}(\omega)T^H(\omega), \tag{6.1}$$

where T^H denotes the Hermitian conjugate of T and $\omega = (\omega_x, \omega_y)^T$ the vector-valued frequency.

We adopt the expressions for the cross spectral densities $\Phi_{c_\mu c_\nu}$ from [11]

$$\Phi_{c_\mu c_\nu}(\omega) = E\left\{e^{-j\omega^T(\Delta_\mu - \Delta_\nu)}\right\}\Phi_{vv}(\omega) + \Phi_{n_\mu n_\nu}(\omega) \qquad (6.2)$$

and assume a normalized power spectrum Φ_{vv} with $\sigma_v^2 = 1$ that corresponds to an exponentially decaying isotropic autocorrelation function with a correlation coefficient between horizontally and vertically adjacent pixels of $\rho_v = 0.93$.

For the k-th displacement error Δ_k, a 2-D normal distribution with variance σ_Δ^2 and zero mean is assumed where the x- and y-components are statistically independent. The expected value in (6.2) depends on the variance of the displacement error with respect to the reference picture c_0 (*absolute displacement accuracy*) and the variance of the difference displacement error between pairs of non-reference pictures (*relative displacement accuracy*). We assume that each picture in a GOP can be the reference picture c_0. That is, there is no preference among the pictures in a GOP and the variances of the absolute displacement error are the same for all $K-1$ motion-compensated pictures. Based on the dyadic decomposition with motion-compensated lifted wavelets and the assumption that there is no preference among the pictures in a GOP, we assume that absolute and relative displacement accuracy are identical. The differences of absolute displacement errors are related to the relative displacement errors as we assume in Sections 6.3.1 and 6.3.2 additive estimated displacements.

$$\Delta_{0j} - \Delta_{0i} = \Delta_{ij} \qquad (6.3)$$

With that, we obtain for the variances of the absolute and relative displacement error components:

$$E\left\{(\Delta_{0j} - \Delta_{0i})^2\right\} = E\left\{\Delta_{ij}^2\right\} \qquad (6.4)$$

$$2\sigma_\Delta^2(1 - \rho_\Delta) = \sigma_\Delta^2 \qquad (6.5)$$

This is only possible with correlated displacement errors such that $\rho_\Delta = 0.5$ [228]. Finally, we abbreviate the expected value in (6.2) with $P(\omega, \sigma_\Delta^2)$ which is the characteristic function of the continuous 2-D Gaussian displacement error.

$$E\left\{e^{-j\omega^T\Delta_k}\right\} := P(\omega, \sigma_\Delta^2) = e^{-\frac{1}{2}\omega^T\omega\sigma_\Delta^2} \qquad (6.6)$$

With that, we obtain for the power spectral density matrix of the motion-compensated pictures

$$
\frac{\Phi_{cc}(\omega)}{\Phi_{vv}(\omega)} = \begin{pmatrix} 1+\alpha(\omega) & P(\omega) & \cdots & P(\omega) \\ P(\omega) & 1+\alpha(\omega) & \cdots & P(\omega) \\ \vdots & \vdots & \ddots & \vdots \\ P(\omega) & P(\omega) & \cdots & 1+\alpha(\omega) \end{pmatrix}. \tag{6.7}
$$

$\alpha = \alpha(\omega)$ is the normalized spectral density of the noise $\Phi_{n_k n_k}(\omega)$ with respect to the spectral density of the "clean" video signal.

$$
\alpha(\omega) = \frac{\Phi_{n_k n_k}(\omega)}{\Phi_{vv}(\omega)} \quad \text{for} \quad k = 0, 1, \ldots, K-1 \tag{6.8}
$$

T represents the dyadic Haar transform or the dyadic 5/3 transform. In terms of decorrelation and coding gain, the 5/3 wavelet performs better than the Haar wavelet as shown in Figs. 6.3 - 6.9. In the following, we are interested in theoretical performance bounds and choose the Karhunen-Loeve Transform (KLT). The normalized eigenvalues of the power spectral density matrix Φ_{cc} are $\lambda_1(\omega) = 1 + \alpha(\omega) + (K-1)P(\omega)$ and $\lambda_{2,3,\ldots,K}(\omega) = 1 + \alpha(\omega) - P(\omega)$. The power spectral density matrix of the transformed signals Φ_{yy} is diagonal.

$$
\frac{\Phi_{yy}(\omega)}{\Phi_{vv}(\omega)} = \begin{pmatrix} \lambda_1(\omega) & 0 & \cdots & 0 \\ 0 & \lambda_2(\omega) & \cdots & 0 \\ \vdots & \vdots & \ddots & \vdots \\ 0 & 0 & \cdots & \lambda_K(\omega) \end{pmatrix} \tag{6.9}
$$

The first eigenvector just adds all components and scales with $1/\sqrt{K}$. For the remaining eigenvectors, any orthonormal basis can be used that is orthogonal to the first eigenvector. That is, the KLT for our signal model is not dependent on ω. Note that for this simple signal model, the Haar transform is also a KLT.

6.3.4   Transform Coding Gain

The rate difference [11] is used to measure the improved compression efficiency for each picture k.

$$
\Delta R_k = \frac{1}{4\pi^2} \int\limits_{-\pi}^{\pi} \int\limits_{-\pi}^{\pi} \frac{1}{2} \log_2 \left(\frac{\Phi_{y_k y_k}(\omega)}{\Phi_{c_k c_k}(\omega)} \right) d\omega \tag{6.10}
$$

It represents the maximum bit-rate reduction (in bit per sample) possible by optimum encoding of the transformed signal y_k, compared to optimum intra-frame encoding of the signal c_k for Gaussian wide-sense stationary signals for the same mean square reconstruction error. A negative ΔR_k corresponds to a reduced bit-rate compared to optimum intra-frame coding. The overall rate difference ΔR is the average over all pictures and is used to evaluate the efficiency of motion-compensated transform coding. Assuming the KLT, we obtain for the overall rate difference

$$\Delta R = \frac{1}{4\pi^2} \int\limits_{-\pi}^{\pi} \int\limits_{-\pi}^{\pi} \frac{K-1}{2K} \log_2 \left(1 - \frac{P(\omega, \sigma_\Delta^2)}{1 + \alpha(\omega)} \right) +$$

$$\frac{1}{2K} \log_2 \left(1 + (K-1) \frac{P(\omega, \sigma_\Delta^2)}{1 + \alpha(\omega)} \right) d\omega. \quad (6.11)$$

The case of a very large number of motion-compensated pictures is of special interest for the comparison to predictive video coding with motion compensation.

$$\Delta R_{K \to \infty} = \frac{1}{4\pi^2} \int\limits_{-\pi}^{\pi} \int\limits_{-\pi}^{\pi} \frac{1}{2} \log_2 \left(1 - \frac{P(\omega, \sigma_\Delta^2)}{1 + \alpha(\omega)} \right) d\omega \quad (6.12)$$

Note that the performance of predictive coding with motion compensation and optimum Wiener filter achieves a rate difference of

$$\Delta R_{\text{MCP}} = \frac{1}{4\pi^2} \int\limits_{-\pi}^{\pi} \int\limits_{-\pi}^{\pi} \frac{1}{2} \log_2 \left(1 - \frac{P^2(\omega, \sigma_\Delta^2)}{[1 + \alpha(\omega)]^2} \right) d\omega. \quad (6.13)$$

We obtain this result from [11], Eqn. 21 with $N = 1$ and $\alpha_0 = \alpha_1 = \alpha$.

Figs. 6.17 and 6.18 depict the rate difference according to (6.11) and (6.13) over the displacement inaccuracy $\beta = \log_2(\sqrt{12}\sigma_\Delta)$ for a residual noise level RNL $= 10 \log_{10}(\sigma_n^2)$ of -100 dB and -30 dB, respectively. Note that the variance of the "clean" video signal $\mathbf{v}$ is normalized to $\sigma_v^2 = 1$. We observe that the rate difference starts to saturate for $K = 32$. This observation is consistent with the experimental results in the previous section. For a very large group of pictures and negligible residual noise, the slope of the rate difference is limited by 1 bit per sample per inaccuracy step, similar to that of predictive coding with motion compensation. Further, transform coding with motion compensation outperforms predictive coding with motion compensation by at most 0.5 bits per sample. For example, if we encode frames with statistically independent additive noise, predictive coding with motion compensation is not capable of predicting the additive noise in the current frame. In this case, prediction

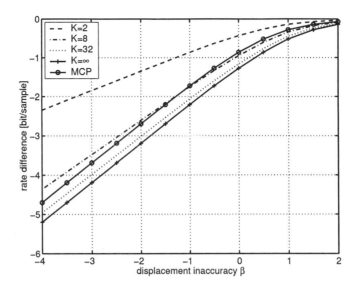

Figure 6.17.   Rate difference for motion-compensated transform coding with groups of K pictures over the displacement inaccuracy β. The performance of predictive coding with motion compensation and Wiener filter is labeled by MCP. The residual noise level is -100 dB.

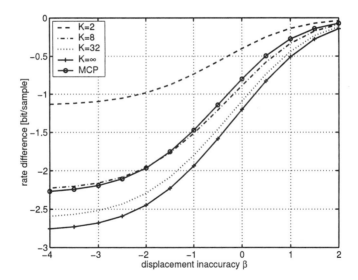

Figure 6.18.   Rate difference for motion-compensated transform coding with groups of K pictures over the displacement inaccuracy β. The performance of predictive coding with motion compensation and Wiener filter is labeled by MCP. The residual noise level is -30 dB.

actually doubles the noise variance in the residual signal and predictive coding performs suboptimally. The advantage of the motion-compensated lifted

5/3 wavelet over motion-compensated predictive coding is also observed in the experimental results.

6.4 Conclusions

This chapter discusses motion compensation for groups of K pictures. We investigate experimentally and theoretically motion-compensated lifted wavelet transforms. The experiments show that the 5/3 wavelet kernel outperforms both the Haar kernel and, in many cases, the reference scheme utilizing motion-compensated predictive coding. The motion-compensated lifted wavelet kernels re-use the motion vectors in the prediction step for the update step by assuming an invertible block-motion field. This assumption seems to be inadequate for sequences with inhomogeneous motion as their rate-distortion performance is weaker than expected.

The theoretical discussion is based on a signal model for K motion-compensated pictures that are decorrelated by a linear transform. The dyadic decomposition of K pictures with motion-compensated lifted wavelets is replaced by an equivalent coding scheme with K motion-compensated pictures and a dyadic wavelet decomposition without motion compensation. That is, we remove the displacement operators in the lifting steps and generate a set of motion-compensated pictures with an additional constraint on the displacement errors. We generalize the model and employ the Karhunen-Loeve Transform to obtain theoretical performance bounds at high bit-rates for motion-compensated 3D transform coding.

The analysis of this model gives the following insights: The coding gain for a group of K pictures is limited and saturates with increasing K. For a very large group of pictures and negligible residual noise, the slope of the rate difference is limited by 1 bit per sample per inaccuracy step. The slope of the rate difference for predictive coding with motion compensation is also limited by 1 bit per sample per inaccuracy step but this coding scheme outperforms predictive coding with motion compensation by up to 0.5 bits per sample. This is also true for very accurate motion compensation when the residual noise dominates the coding gain.

Chapter 7

SUMMARY

This work discusses video coding with superimposed motion-compensated signals. We build on the theory of multihypothesis motion-compensated prediction for video coding and introduce the concept of motion compensation with complementary hypotheses. Multihypothesis motion compensation linearly combines more than one motion-compensated signal to form the superimposed motion-compensated signal. Motion-compensated signals that are used for the superposition are referred to as hypotheses. Further, a displacement error that captures the inaccuracy of motion compensation is associated with each hypothesis. As the accuracy of motion compensation is equal for all hypotheses, the displacement errors are identically distributed.

This work proposes that the multiple displacement errors are jointly distributed and, in particular, correlated. Investigations show that there is no preference among N hypotheses and we conclude that all non-equal pairs of displacement errors are characterized by one correlation coefficient. As the covariance matrix of the jointly distributed displacement errors is nonnegative definite, we can determine the valid range of the displacement error correlation coefficient dependent on the number of superimposed hypotheses.

We investigate the efficiency of superimposed motion compensation as a function of the displacement error correlation coefficient. We observe that decreasing the displacement error correlation coefficient improves the efficiency of superimposed motion compensation. We conclude that motion compensation with complementary hypotheses results in maximally negatively correlated displacement error.

Motion compensation with complementary hypotheses implies two major results for the efficiency of superimposed motion-compensated prediction:

First, the slope of the rate difference reaches up to 2 bits per sample per motion inaccuracy step whereas for single hypothesis motion-compensated prediction this slope is limited to 1 bit per sample per motion inaccuracy step. Here, we measure the rate difference with respect to optimum intra-frame encoding and use a high-rate approximation. Second, this slope of 2 bits per sample per inaccuracy step is already achieved for $N = 2$ complementary hypotheses. If we just average the hypotheses, the performance converges at constant slope to that of a very large number of hypotheses. That is, the largest portion of the achievable gain is already accomplished with $N = 2$ complementary hypotheses. If we employ the optimum Wiener filter, the coding performance improves for doubling the number of complementary hypotheses by at least 0.5 bits per sample at constant slope.

In this work, we investigate motion compensation with complementary hypotheses by integrating superimposed motion-compensated prediction into the ITU-T Rec. H.263. We linearly combine up to 4 motion-compensated blocks chosen from up to 20 previous reference frames to improve the performance of inter-predicted pictures. To determine the best N-hypothesis for each predicted block, we utilize an iterative algorithm that improves successively conditional optimal hypotheses. Our experiments show that superimposed prediction works efficiently for both 16×16 and 8×8 blocks. Multiple reference frames enhance the efficiency of superimposed prediction. The superposition gain and the multiframe gain do not only add up; superimposed prediction benefits from hypotheses which can be chosen from several reference frames. Superimposed prediction with two hypotheses and ten reference frames achieves coding gains up to 2.7 dB, or equivalently, bit-rate savings up to 30% for the sequence *Mobile & Calendar* when compared to the one-hypothesis reference codec with one reference frame.

We explore theoretically why superimposed prediction benefits from multiframe motion compensation. We model multiframe motion compensation by forward-adaptive hypothesis switching and show that switching among M hypotheses with statistically independent displacement error reduces the displacement error variance by up to a factor of M. For motion-compensated prediction, we obtain the following performance bounds: Doubling the number of reference pictures for single hypothesis prediction reduces the bit-rate of the residual encoder by at most 0.5 bits per sample. Whereas doubling the number of reference pictures for prediction with complementary hypotheses reduces the bit-rate of the residual encoder by at most 1 bit per sample.

This work also investigates motion compensation with complementary hypotheses for B-pictures in the emerging ITU-T Rec. H.264. We focus on reference picture selection and linearly combined motion-compensated prediction signals. We show that bidirectional prediction exploits partially the efficiency of combined prediction signals. Superimposed prediction chooses hypotheses from an arbitrary set of reference pictures and, by this, outperforms bidirectional prediction. That is, superimposed motion-compensated prediction with multiple reference frames allows a more general form of B-pictures. In addition to the generalization of the bidirectional mode, we allow that previously decoded B-pictures can be reference pictures for other B-pictures. Again, we observe that multiframe motion compensation enhances the efficiency of superimposed prediction for hybrid video coding.

Finally, we discuss superimposed motion-compensated signals for motion-compensated 3D subband coding of video. We investigate experimentally and theoretically motion-compensated lifted wavelet transforms for the temporal subband decomposition. The experiments show that the 5/3 wavelet kernel outperforms both the Haar kernel and, in many cases, the reference scheme utilizing motion-compensated predictive coding. Based on the motion-compensated lifting scheme, we develop an analytical model describing motion compensation for groups of K pictures.

The theoretical discussion is based on a signal model for K motion-compensated pictures that are decorrelated by a linear transform. The dyadic decomposition of K pictures with motion-compensated lifted wavelets is replaced by an equivalent coding scheme with K motion-compensated pictures and a dyadic wavelet decomposition without motion compensation. That is, we remove the displacement operators in the lifting steps and generate a set of motion-compensated pictures with an additional constraint on the displacement errors. We generalize the model and employ the Karhunen-Loeve Transform to obtain theoretical performance bounds at high bit-rates for motion-compensated 3D transform coding.

The analysis of this model gives the following insights: The coding gain for a group of K pictures is limited and starts to saturate for $K = 32$ pictures. For a very large group of pictures and negligible residual noise, the slope of the rate difference is limited by 1 bit per sample per inaccuracy step. The slope of the rate difference for motion-compensated prediction is also limited by 1 bit per sample per inaccuracy step but this coding scheme outperforms motion-compensated prediction by at most 0.5 bits per sample. This is also true for very accurate motion compensation when the residual noise dominates the coding gain.

In summary, we characterize the relation between motion-compensated signals and, depending on this relation, investigate their efficiency for video compression.

Appendix A
Mathematical Results

A.1 Singularities of the Displacement Error Covariance Matrix

The prediction scheme shows no preferences among the individual hypotheses. This is reflected in the symmetry of the displacement error covariance matrix. The variance of the displacement error for each hypothesis is σ_Δ^2 and the correlation coefficient between any hypothesis pair is ρ_Δ. With that, the covariance matrix of the displacement error vector reads

$$C - \sigma_\Delta^2 \begin{pmatrix} 1 & \rho_\Delta & \cdots & \rho_\Delta \\ \rho_\Delta & 1 & \cdots & \rho_\Delta \\ \vdots & \vdots & \ddots & \vdots \\ \rho_\Delta & \rho_\Delta & \cdots & 1 \end{pmatrix}. \tag{A.1}$$

In order to determine the singularities of the matrix, we decompose the covariance matrix into the identity matrix I and the matrix $\mathbf{1}\mathbf{1}^T$ with each element equal to one.

$$C = \sigma_\Delta^2 \rho_\Delta \mathbf{1}\mathbf{1}^T - \sigma_\Delta^2 (\rho_\Delta - 1)I \tag{A.2}$$

The covariance matrix is singular if its determinant is zero

$$\det\left(\mathbf{1}\mathbf{1}^T - \frac{\rho_\Delta - 1}{\rho_\Delta} I\right) = 0. \tag{A.3}$$

The eigenvalues of the matrix $\mathbf{1}\mathbf{1}^T$ are $\lambda = \{N, 0\}$ which can also be obtained by solving $\det(\mathbf{1}\mathbf{1}^T - \lambda I) = 0$. We obtain two singularities for the covariance matrix:

$$\rho_\Delta = \frac{1}{1 - N} \quad \text{and} \quad \rho_\Delta = 1. \tag{A.4}$$

A.2   A Class of Matrices and their Eigenvalues

We consider the following class of normalized $N \times N$ matrices with the parameter $a \in \mathcal{R}$:

$$C = \begin{pmatrix} 1 & a & \cdots & a \\ a & 1 & \cdots & a \\ \vdots & \vdots & \ddots & \vdots \\ a & a & \cdots & 1 \end{pmatrix} \tag{A.5}$$

In order to determine the eigenvalues of C, we decompose the matrix into the identity matrix I and the matrix $\mathbf{1}\mathbf{1}^T$ with each element equal to one.

$$C = a\mathbf{1}\mathbf{1}^T + (1-a)I \tag{A.6}$$

The eigenvalues λ_i of C solve $\det(C - \lambda_i I) = 0$ and with (A.6), we can write for $a \neq 0$

$$\det\left(\mathbf{1}\mathbf{1}^T - \frac{\lambda_i - 1 + a}{a}I\right) = 0. \tag{A.7}$$

The eigenvalues of the matrix $\mathbf{1}\mathbf{1}^T$ are $\lambda_1 = N$, 1-fold, and $\lambda_2 = 0$, $(N-1)$-fold. Both can be obtained by solving $\det(\mathbf{1}\mathbf{1}^T - \lambda_i I) = 0$. With $a \neq 0$, we have the following eigenvalues for C:

$$\begin{aligned} \lambda_1 &= 1 + (N-1)a & \text{1-fold} \\ \lambda_2 &= 1 - a & (N-1)\text{-fold} \end{aligned} \tag{A.8}$$

If $a = 0$, the N-fold eigenvalue of C is 1.

A.3   Inverse of the Power Spectral Density Matrix

The optimum Wiener filter requires the inverse of the hypothesis power spectral density matrix $\Phi_{\mathbf{cc}}(\omega)$. After normalization, the matrix H_1 of the form

$$H_1 = \begin{pmatrix} c & 1 & \cdots & 1 \\ 1 & c & \cdots & 1 \\ \vdots & \vdots & \ddots & \vdots \\ 1 & 1 & \cdots & c \end{pmatrix} \tag{A.9}$$

is inverted. In the following, we prefer a representation of H_1 with the vector $\mathbf{1} = (1, 1, \ldots, 1)^T$ and the identity matrix I

$$H_1 = \mathbf{1}\mathbf{1}^T + (c-1)I. \tag{A.10}$$

Due to the symmetry of H_1, its inverse shows the same structure

$$H_1^{-1} = k\left[\mathbf{1}\mathbf{1}^T + (d-1)I\right], \tag{A.11}$$

where k and d are scalar constants. As the inverse of the non-singular matrix ($c \neq 1, c \neq 1-N$) is unique, we obtain

$$\begin{aligned} k &= \frac{1}{(c-1)(1-c-N)}, \tag{A.12} \\ d &= 2 - c - N, \tag{A.13} \end{aligned}$$

and for the inverse

$$H_1^{-1} = \frac{11^T + (1 - c - N)I}{(c - 1)(1 - c - N)}. \tag{A.14}$$

In the case that the noise variance for each hypothesis is not identical, a normalized matrix

$$H_2 = \begin{pmatrix} c_1 & a_1 a_2 & \cdots & a_1 a_N \\ a_2 a_1 & c_2 & \cdots & a_2 a_N \\ \vdots & \vdots & \ddots & \vdots \\ a_N a_1 & a_N a_2 & \cdots & c_N \end{pmatrix} \tag{A.15}$$

is inverted. A representation of H_2 with the vector $a = (a_1, a_2, \ldots, a_N)^T$ and the diagonal matrix diag($\cdot$) reads

$$H_2 = aa^T + \mathrm{diag}(c_i - a_i^2). \tag{A.16}$$

The inverse shows the same symmetry as H_2:

$$H_2^{-1} = k \left[bb^T + \mathrm{diag}(d_i - b_i^2) \right], \tag{A.17}$$

where k is a scalar constant and b and d are vectors of size N. As the inverse of a non-singular matrix is unique, the parameters of H_2^{-1} can be calculated for $a_i^2 \neq c_i$, $i = 1, 2, \ldots, N$, and $k \neq 0$ with

$$k = \frac{1}{\displaystyle\sum_{j=1}^{N} \frac{a_j^2}{a_j^2 - c_j} - 1}, \tag{A.18}$$

$$b_i = \frac{a_i}{a_i^2 - c_i}, \tag{A.19}$$

$$d_i = \frac{1}{a_i^2 - c_i} \left[\frac{a_i^2}{a_i^2 - c_i} - \frac{1}{k} \right]. \tag{A.20}$$

A simplified expression for the inverse reads

$$H_2^{-1} = \frac{bb^T}{\displaystyle\sum_{j=1}^{N} \frac{a_j^2}{a_j^2 - c_j} - 1} - \mathrm{diag}\left(\frac{1}{a_i^2 - c_i} \right) \quad \text{with} \quad b_i = \frac{a_i}{a_i^2 - c_i}. \tag{A.21}$$

This solution for H_2^{-1} is only valid for $a_i^2 \neq c_i$ and $k \neq 0$.

The special case $a = 1$ is also of interest for the hypothesis power spectrum matrix. H_3 reads

$$H_3 = \begin{pmatrix} c_1 & 1 & \cdots & 1 \\ 1 & c_2 & \cdots & 1 \\ \vdots & \vdots & \ddots & \vdots \\ 1 & 1 & \cdots & c_N \end{pmatrix} \tag{A.22}$$

and its inverse can be calculated by

$$H_3^{-1} = \frac{bb^T}{\displaystyle\sum_{j=1}^{N} \frac{1}{1 - c_j} - 1} - \mathrm{diag}\left(\frac{1}{1 - c_i} \right) \quad \text{with} \quad b_i = \frac{1}{1 - c_i}. \tag{A.23}$$

This solution holds only for $c_i \neq 1$, $i = 1, 2, \ldots, N$, and $\sum_{j=1}^{N} \frac{1}{1-c_j} \neq 1$.

A.4   Power Spectral Density of a Frame

The 2D continuous-space Fourier transform $\mathcal{F}\{\cdot\}$ of the 2D continuous signal $v(x, y)$ in Cartesian coordinates with $x, y \in \mathcal{R}$ is defined by

$$V(\omega_x, \omega_y) = \int_{\mathcal{R}^2} v(x, y) e^{-j(\omega_x x + \omega_y y)} dx dy. \tag{A.24}$$

The 2D continuous-space Fourier transform of the 2D continuous signal $\overset{\circ}{v}(r, \theta)$ in cylindrical coordinates is given by

$$\overset{\circ}{V}(\omega_r, \omega_\theta) = \int_0^{2\pi} \int_0^\infty \overset{\circ}{v}(r, \theta) e^{-jr\omega_r \cos(\theta - \omega_\theta)} r \, dr \, d\theta \tag{A.25}$$

where

$$x = r \cos\theta, \qquad\qquad \omega_x = \omega_r \cos\omega_\theta, \tag{A.26}$$

$$y = r \sin\theta, \qquad\qquad \omega_y = \omega_r \sin\omega_\theta. \tag{A.27}$$

The Fourier transform of the isotropic, exponentially decaying, space-continuous function

$$\overset{\circ}{v}(r, \theta) = e^{-\omega_0 r} \quad \forall\theta \tag{A.28}$$

with $\omega_0 > 0$ is also isotropic.

$$\overset{\circ}{V}(\omega_r, \omega_\theta) = \int_0^{2\pi} \int_0^\infty e^{-[\omega_0 + j\omega_r \cos(\theta)]r} r \, dr \, d\theta \tag{A.29}$$

$$= \int_0^{2\pi} \frac{1}{[\omega_0 + j\omega_r \cos(\theta)]^2} d\theta \tag{A.30}$$

$$= \frac{1}{\omega_0^2} \int_0^{2\pi} \frac{1}{[1 + j\Omega \cos(\theta)]^2} d\theta, \quad \Omega = \frac{\omega_r}{\omega_0} \tag{A.31}$$

$$= \frac{2j}{\omega_0^2} \int_{1+j\Omega}^{1-j\Omega} \frac{1}{x^2 \sqrt{x^2 - 2x + 1 + \Omega^2}} dx, \quad x = 1 + j\Omega \cos(\theta) \tag{A.32}$$

$$= \frac{2\pi}{\omega_0^2} \left(1 + \Omega^2\right)^{-\frac{3}{2}} \tag{A.33}$$

$$= \frac{2\pi}{\omega_0^2} \left(1 + \frac{\omega_r^2}{\omega_0^2}\right)^{-\frac{3}{2}} \tag{A.34}$$

The integral in (A.32) is solved in [229, 230]. In Cartesian coordinates, the Fourier transform of the function $v(x, y) = e^{-\omega_0 \sqrt{x^2 + y^2}}$ reads

$$V(\omega_x, \omega_y) = \frac{2\pi}{\omega_0^2} \left(1 + \frac{\omega_x^2 + \omega_y^2}{\omega_0^2}\right)^{-\frac{3}{2}}. \tag{A.35}$$

Glossary

Acronyms

BL	Baseline
CABAC	Context-Based Adaptive Binary Arithmetic Coding
CIF	Common Intermediate Format
DCT	Discrete Cosine Transform
DHT	Dyadic Haar Transform
D5/3T	Dyadic 5/3 Transform
ECVQ	Entropy Constrained Vector Quantization
GOP	Group of Pictures
HP	Half-Pel
INTER4V	Inter-prediction mode with four 8×8 blocks
INTER2H	Inter-prediction mode with two hypotheses
INTER4H	Inter-prediction mode with four hypotheses
INTER4VMH	Inter-prediction mode with four 8×8 superimposed blocks
IP	Integer-Pel
ITU	International Telecommunication Union
ITU-T	ITU Telecommunication Standardization Sector
KLT	Karhunen-Loeve Transform
MCP	Motion-Compensated Prediction
MHP	Superimposed Prediction
MPEG	Moving Picture Experts Group

OBMC	Overlapped Block Motion Compensation
PDF	Probability Density Function
PSNR	Peak Signal to Noise Ratio
QCIF	Quarter Common Intermediate Format
RNL	Residual Noise Level
TML-9	H.26L Test Model Software Version 9
UVLC	Universal Variable Length Coding
VBS	Variable Block Size
VCEG	Video Coding Experts Group
VQ	Vector Quantization

Probability Theory

$\mathcal{R}$	Set of real numbers
Π	Two-dimensional orthogonal unit grid
$\mathbf{a}$	Random variable, process, or field
$\Pr\{\mathbf{a} \leq a\}$	Probability of the event $\{\mathbf{a} \leq a\}$
$R_{\mathbf{a}}(a)$	Reliability function of $\mathbf{a}$
$p_{\mathbf{a}}(a)$	PDF of $\mathbf{a}$
$\sigma_{\mathbf{a}}^2$	Variance of $\mathbf{a}$
$\rho_{\mathbf{a}}$	Correlation coefficient associated with $\mathbf{a}$
$\phi_{\mathbf{ab}}(\cdot)$	Correlation function between $\mathbf{a}$ and $\mathbf{b}$
$\Phi_{\mathbf{ab}}(\cdot)$	Cross spectral density of $\mathbf{a}$ and $\mathbf{b}$
$E\{\cdot\}$	Expectation operator

Matrix Algebra

$\mathbf{1}$	Column vector with all entries equal to one
$\|\cdot\|_2^2$	Square norm of a vector
$\|\cdot\|$	Length of a code word vector
$\det(\cdot)$	Determinant of a matrix
$\mathrm{diag}(\cdot)$	Diagonal matrix
I	Identity matrix
$.^*$	Complex conjugate operator

$.^H$	Hermitian conjugate operator
$.^T$	Transpose operator

Transforms

$\mathcal{F}_* \{\cdot\}$	2D band-limited discrete-space Fourier transform
$\mathcal{F} \{\cdot\}$	2D continuous-space Fourier transform

References

[1] B. Girod, "Motion-compensating prediction with fractional-pel accuracy", *IEEE Transactions on Communications*, vol. 41, no. 4, pp. 604–612, Apr. 1993.

[2] G.J. Sullivan and R.L. Baker, "Rate-distortion optimized motion compensation for video compression using fixed or variable size blocks", in *Proceedings of the IEEE Global Telecommunications Conference*, Phoenix, AZ, Dec. 1991, vol. 3, pp. 85–90.

[3] M. Gothe and J. Vaisey, "Improving motion compensation using multiple temporal frames", in *Proceedings of the IEEE Pacific Rim Conference on Communications, Computers, and Signal Processing*, Victoria, BC, May 1993, pp. 157–160.

[4] M. Budagavi and J.D. Gibson, "Multiframe block motion compensated video coding for wireless channels", in *Thirtieth Asilomar Conference on Signals, Systems and Computers*, Nov. 1996, vol. 2, pp. 953–957.

[5] T. Wiegand, X. Zhang, and B. Girod, "Block-based hybrid video coding using motion-compensated long-term memory prediction", in *Proceedings of the Picture Coding Symposium*, Berlin, Germany, Sept. 1997, pp. 153–158.

[6] G.J. Sullivan, "Multi-hypothesis motion compensation for low bit-rate video coding", in *Proceedings of the IEEE International Conference on Acoustics, Speech and Signal Processing*, Minneapolis, MN, Apr. 1993, vol. 5, pp. 437–440.

[7] S. Nogaki and M. Ohta, "An overlapped block motion compensation for high quality motion picture coding", in *Proceedings of the IEEE International Symposium on Circuits and Systems*, May 1992, pp. 184–187.

[8] S. Ericsson, "Fixed and adaptive predictors for hybrid predictive/transform coding", *IEEE Transactions on Communications*, vol. 33, no. 12, pp. 1291–1302, Dec. 1985.

[9] H.G. Musmann, P. Pirsch, and H.J. Grallert, "Advances in picture coding", *Proceedings of the IEEE*, vol. 73, no. 4, pp. 523–548, Apr. 1985.

[10] A. Puri, R. Aravind, B.G. Haskell, and R. Leonardi, "Video coding with motion-compensated interpolation for CD-ROM applications", *Signal Processing: Image Communication*, vol. 2, no. 2, pp. 127–144, Aug. 1990.

[11] B. Girod, "Efficiency analysis of multihypothesis motion-compensated prediction for video coding", *IEEE Transactions on Image Processing*, vol. 9, no. 2, pp. 173–183, Feb. 2000.

[12] B. Girod, "The efficiency of motion-compensating prediction for hybrid coding of video sequences", *IEEE Journal on Selected Areas in Communications*, vol. SAC-5, no. 7, pp. 1140–1154, Aug. 1987.

[13] ISO/IEC, *13818-2 Information Technology - Generic Coding of Moving Pictures and Associated Audio Information: Video (MPEG-2)*, 1996.

[14] ITU-T, *Recommendation H.263++ (Video Coding for Low Bitrate Communication)*, 2000.

[15] M. Flierl, "Bewegungskompensierte Multihypothesen-Langzeitprädiktion", Diploma thesis, Telecommunications Laboratory, University of Erlangen-Nuremberg, Erlangen, Germany, Sept. 1997.

[16] M. Flierl, T. Wiegand, and B. Girod, "A locally optimal design algorithm for block-based multi-hypothesis motion-compensated prediction", in *Proceedings of the Data Compression Conference*, Snowbird, Utah, Apr. 1998, pp. 239–248.

[17] S.-W. Wu and A. Gersho, "Joint estimation of forward and backward motion vectors for interpolative prediction of video", *IEEE Transactions on Image Processing*, vol. 3, no. 5, pp. 684–687, Sept. 1994.

[18] S.C. Knauer, "Real-time video compression algorithm for Hadamard transform processing", *IEEE Transactions on Electromagnetic Compatibility*, vol. EMC-18, pp. 28–36, Feb. 1976.

[19] T. Kronander, "Motion compensated 3-dimensional wave-form image coding", in *Proceedings of the IEEE International Conference on Acoustics, Speech and Signal Processing*, Glasgow, Scotland, May 1989, vol. 3, pp. 1921–1924.

[20] A. Secker and D. Taubman, "Motion-compensated highly scalable video compression using an adaptive 3D wavelet transform based on lifting", in *Proceedings of the IEEE International Conference on Image Processing*, Thessaloniki, Greece, Oct. 2001, vol. 2, pp. 1029–1032.

[21] G. Côté, B. Erol, M. Gallant, and F. Kossentini, "H.263+: Video coding at low bit rates", *IEEE Transactions on Circuits and Systems for Video Technology*, vol. 8, no. 7, pp. 849–866, Nov. 1998.

[22] ITU-T Video Coding Experts Group and ISO/IEC Moving Picture Experts Group, *Working Draft Number 2, Revision 7*, Apr. 2002, ftp:// ftp.imtc-files.org/ jvt-experts/ draft_standard/ jwd2r7.zip.

[23] D. Taubman and A. Zakhor, "Multirate 3-d subband coding of video", *IEEE Transactions on Image Processing*, vol. 3, no. 5, pp. 572–588, Sept. 1994.

[24] J.-R. Ohm, "Three-dimensional subband coding with motion compensation", *IEEE Transactions on Circuits and Systems for Video Technology*, vol. 3, no. 5, pp. 559–571, Sept. 1994.

[25] S.-J. Choi and J.W. Woods, "Motion-compensated 3-d subband coding of video", *IEEE Transactions on Image Processing*, vol. 8, no. 2, pp. 155–167, Feb. 1999.

[26] J.R. Jain and A.K. Jain, "Displacement measurement and its application in interframe image coding", *IEEE Transactions on Communications*, vol. 29, no. 12, pp. 1799–1808, Dec. 1981.

[27] Y. Ninomiya and Y. Ohtsuka, "A motion-compensated interframe coding scheme for television pictures", *IEEE Transactions on Communications*, vol. COM-30, no. 1, pp. 201–211, Jan. 1982.

[28] Atul Puri, *Efficient Motion-Compensated Coding for Low Bit-Rate Video Applications*, PhD thesis, The City University of New York, 1988.

[29] Y. Nakaya and H. Harashima, "Motion compensation based on spatial transformations", *IEEE Transactions on Circuits and Systems for Video Technology*, vol. 4, no. 3, pp. 339–356, June 1994.

[30] H. Li and R. Forchheimer, "A transformed block-based motion compensation technique", *IEEE Transactions on Communications*, vol. 43, no. 2/3/4, pp. 1673–1676, Feb. 1995.

[31] Y. Yokoyama, Y. Miyamoto, and M. Ohta, "Very low bit rate video coding using arbitrarily shaped region-based motion compensation", *IEEE Transactions on Circuits and Systems for Video Technology*, vol. 5, no. 6, pp. 500–507, Dec. 1995.

[32] E. Dubois and S. Sabri, "Noise reduction in image sequences using motion-compensated temporal filtering", *IEEE Transactions on Communications*, vol. COM-32, no. 7, pp. 826–831, July 1984.

[33] J. Kim and J.W. Woods, "Spatio-temporal adaptive 3-d kalman filter for video", *IEEE Transactions on Image Processing*, vol. 6, no. 3, pp. 414–424, Mar. 1997.

[34] J. Kim and J.W. Woods, "3-d kalman filter for image motion estimation", *IEEE Transactions on Image Processing*, vol. 7, no. 1, pp. 42–52, Jan. 1998.

[35] A.N. Netravali and J.D. Robbins, "Video signal interpolation using motion estimation", United States Patent 4 383 272, May 1983, Filed: April 13, 1981.

[36] M. Soryani and R.J. Clarke, "Image segmentation and motion-adaptive frame interpolation for coding moving sequences", in *Proceedings of the IEEE International Conference on Acoustics, Speech and Signal Processing*, Glasgow, Scotland, May 1989, pp. 1882–1885.

[37] R. Thoma and M. Bierling, "Motion compensating interpolation considering covered and uncovered background", *Signal Processing: Image Communication*, vol. 1, no. 2, pp. 191–212, Oct. 1989.

[38] C. Cafforio, F. Rocca, and S. Tubaro, "Motion compensated image interpolation", *IEEE Transactions on Communications*, vol. 38, no. 2, pp. 215–222, Feb. 1990.

[39] J. Ribas-Corbera and J. Sklansky, "Interpolation of cinematic sequences", in *Proceedings of the IEEE Workshop on Applications of Computer Vision*, Nov. 1992, pp. 36–44.

[40] L. Zhao and Z. Zhou, "A new algorithm for motion-compensated frame interpolation", in *Proceedings of the IEEE International Symposium on Circuits and Systems*, Chicago, Illinois, May 1993, pp. 9–12.

[41] R. Castagno, "A method for motion adaptive frame rate up-conversion", *IEEE Transactions on Circuits and Systems for Video Technology*, vol. 6, no. 5, pp. 436–446, Oct. 1996.

[42] E. Francois, J.-F. Vial, and B. Chupeau, "Coding algorithm with region-based motion compensation", *IEEE Transactions on Circuits and Systems for Video Technology*, vol. 7, no. 1, pp. 97–108, Feb. 1997.

[43] J. Kovačević, R.J. Safranek, and E.M. Yeh, "Adaptive bidirectional time-recursive interpolation for deinterlacing", in *Proceedings of the Data Compression Conference*, Snowbird, UT, Mar. 1995, p. 446.

[44] J. Kovačević, R.J. Safranek, and E.M. Yeh, "Deinterlacing by successive approximation", *IEEE Transactions on Image Processing*, vol. 6, no. 2, pp. 339–344, Feb. 1997.

[45] C.-K. Wong and O.C. Au, "Modified motion compensated temporal frame interpolation for very low bit rate video", in *Proceedings of the IEEE International Conference on Acoustics, Speech and Signal Processing*, Atlanta, GA, May 1996, vol. 4, pp. 2327–2330.

[46] C.-W. Tang and O.C. Au, "Comparison between block-based and pixel-based temporal interpolation for video coding", in *Proceedings of the IEEE International Symposium on Circuits and Systems*, Monterey, CA, June 1998, pp. 122–125.

[47] C.-K. Wong, O.C. Au, and C.-W. Tang, "Motion compensated temporal interpolation with overlapping", in *Proceedings of the IEEE International Symposium on Circuits and Systems*, Atlanta, GA, May 1996, pp. 608–611.

[48] C.-W. Tang and O.C. Au, "Unidirectional motion compensated temporal interpolation", in *Proceedings of the IEEE International Symposium on Circuits and Systems*, Hong Kong, June 1997, pp. 1444–1447.

[49] A.M. Tourapis, H.-Y. Cheong, M.L. Liou, and O.C. Au, "Temporal interpolation of video sequences using zonal based algorithms", in *Proceedings of the IEEE International Conference on Image Processing*, Thessaloniki, Greece, Oct. 2001, pp. 895–898.

[50] A.M. Tourapis, H.-Y. Cheong, O.C. Au, and M.L. Liou, "N-dimensional zonal algorithms. the future of block based motion estimation?", in *Proceedings of the IEEE International Conference on Image Processing*, Thessaloniki, Greece, Oct. 2001, pp. 506–509.

[51] A.M. Tourapis, O.C. Au, and M.L. Liou, "Highly efficient predictive zonal algorithms for fast block-matching motion estimation", *IEEE Transactions on Circuits and Systems for Video Technology*, vol. 12, no. 10, pp. 934–947, Oct. 2002.

[52] T. Micke, "Vergleich eines prädiktiven und eines interpolativen bewegungskompensierenden Codierverfahrens für Fersehbildsignale", Diploma thesis, Institut für Theoretische Nachrichtentechnik und Informationsverarbeitung, Universität Hannover, Hannover, Germany, 1986.

[53] B.G. Haskell and A. Puri, "Conditional motion compensated interpolation of digital motion video", United States Patent 4 958 226, Sept. 1990, Filed: September 27, 1989.

[54] J. Yonemitsu and B.D. Andrews, "Video signal coding method", United States Patent 5 155 593, Oct. 1992, Filed: September 27, 1990.

[55] A. Puri and R. Aravind, "On comparing motion-interpolating structures for video coding", in *Proceedings of the SPIE Conference on Visual Communications and Image Processing*, Lausanne, Switzerland, Oct. 1990, pp. 1560–1571.

[56] A. Puri and R. Aravind, "Motion-compensated video coding with adaptive perceptual quantization", *IEEE Transactions on Circuits and Systems for Video Technology*, vol. 1, no. 4, pp. 351–361, Dec. 1991.

[57] A. Puri, R. Aravind, and B. Haskell, "Adaptive frame/field motion compensated video coding", *Signal Processing: Image Communication*, vol. 5, no. 1-2, pp. 39–58, Feb. 1993.

[58] A. Puri, L. Yan, and B.G. Haskell, "Temporal resolution scalable video coding", in *Proceedings of the IEEE International Conference on Image Processing*, Austin, TX, Nov. 1994, pp. 947–951.

[59] T. Shanableh and M. Ghanbari, "The improtance of the bi-directionally predicted pictures in video streaming", *IEEE Transactions on Circuits and Systems for Video Technology*, vol. 11, no. 3, pp. 402–414, Mar. 2001.

[60] W.E. Lynch, "Bidirectional motion estimation based on P frame motion vectors and area overlap", in *Proceedings of the IEEE International Conference on Acoustics, Speech and Signal Processing*, San Francisco, CA, Mar. 1992, vol. 3, pp. 445–448.

[61] R. Krishnamurthy, J.W. Woods, and P. Moulin, "Frame interpolation and bidirectional prediction of video using compactly encoded optical-flow fields and label fields", *IEEE Transactions on Circuits and Systems for Video Technology*, vol. 9, no. 5, pp. 713–726, Aug. 1999.

[62] A.M. Tekalp, *Digital Video Processing*, Prentice Hall, London, 1995.

[63] ITU-T, *Recommendation H.263, Version 2 (Video Coding for Low Bitrate Communication)*, 1998.

[64] Siu-Wai Wu, *Enhanced Image and Video Compression with Constraints on the Bit Stream Format*, PhD thesis, University of California, Santa Barbara, Mar. 1993.

[65] M.-K. Kim and J.-K. Kim, "Efficient motion estimation algorithm for bidirectional prediction scheme", *IEE Electronics Letters*, vol. 30, no. 8, pp. 632–633, Apr. 1994.

[66] G.J. Sullivan and M.T. Orchard, "Methods of reduced-comlexity overlapped block motion compensation", in *Proceedings of the IEEE International Conference on Image Processing*, Austin, TX, Nov. 1994, vol. 2, pp. 957–961.

[67] M.T. Orchard and G.J. Sullivan, "Overlapped block motion compensation: An estimation-theoretic approach", *IEEE Transactions on Image Processing*, vol. 3, no. 5, pp. 693–699, Sept. 1994.

[68] J. Katto and M. Ohta, "An analytical framework for overlapped motion compensation", in *Proceedings of the IEEE International Conference on Acoustics, Speech and Signal Processing*, 1995, vol. 5, pp. 2189–2192.

[69] G.J. Sullivan and R.L. Baker, "Motion compensation for video compression using control grid interpolation", in *Proceedings of the IEEE International Conference on Acoustics, Speech and Signal Processing*, Toronto, Canada, May 1991, pp. 2713–2716.

[70] H. Watanabe and S. Singhal, "Windowed motion compensation", in *Proceedings of the SPIE Conference on Visual Communications and Image Processing*, Boston, MA, Nov. 1991, pp. 582–589.

[71] B. Tao and M.T. Orchard, "Removal of motion uncertainty and quantization noise in motion compensation", *IEEE Transactions on Circuits and Systems for Video Technology*, vol. 11, no. 1, pp. 80–90, Jan. 2001.

[72] B. Tao, M.T. Orchard, and B. Dickinson, "Joint application of overlapped block motion compensation and loop filtering for low bit-rate video coding", in *Proceedings of the IEEE International Conference on Image Processing*, Santa Barbara, CA, Oct. 1997, pp. 626–629.

[73] B. Tao and M.T. Orchard, "Window design for overlapped block motion compensation through statistical motion modeling", in *Thirty-First Asilomar Conference on Signals, Systems, and Computers*, Nov. 1997, vol. 1, pp. 372–376.

[74] B. Tao and M.T. Orchard, "Non-iterative motion estimation for overlapped block motion compensation", in *Proceedings of the SPIE Conference on Visual Communications and Image Processing*, San Jose, CA, Jan. 1998, pp. 1032–1040.

[75] B. Tao and M.T. Orchard, "Gradient-based residual variance modeling and its applications to motion-compensated video coding", *IEEE Transactions on Image Processing*, vol. 10, no. 1, pp. 24–35, Jan. 2001.

[76] B. Tao and M.T. Orchard, "A parametric solution for optimal overlapped block motion compensation", *IEEE Transactions on Image Processing*, vol. 10, no. 3, pp. 341–350, Mar. 2001.

[77] P. Strobach, "Tree-structured scene adaptive coder", *IEEE Transactions on Communications*, vol. 38, no. 4, pp. 477–486, Apr. 1990.

[78] G.J. Sullivan and R.L. Baker, "Efficient quadtree coding of images and video", *IEEE Transactions on Image Processing*, vol. 3, pp. 327–331, May 1994.

[79] J. Lee, "Optimal quadtree for variable block size motion estimation", in *Proceedings of the IEEE International Conference on Image Processing*, Washington, DC, Oct. 1995, pp. 480–483.

[80] J. Ribas-Corbera and D. L. Neuhoff, "On the optimal block size for block-based, motion-compensated video coders", in *Proceedings of the SPIE Conference on Visual Communications and Image Processing*, San Jose, CA, Jan. 1997, vol. 2, pp. 1132–1143.

[81] M. Flierl, "Untersuchung von Algorithmen zur Bestimmung von Quadtree-Strukturen bei der Videocodierung mit Blöcken variabler Größe", Project thesis, Telecommunications Laboratory, University of Erlangen-Nuremberg, Erlangen, Germany, Mar. 1997.

[82] T. Wiegand, M. Flierl, and B. Girod, "Entropy-constrained design of quadtree video coding schemes", in *Proceedings of the International Conference on Image Processing and its Applications*, Dublin, Ireland, July 1997, vol. 1, pp. 36–40.

[83] N. Mukawa and H. Kuroda, "Uncovered background prediction in interframe coding", *IEEE Transactions on Communications*, vol. COM-33, no. 11, pp. 1227–1231, Nov. 1985.

[84] D. Hepper, "Efficiency analysis and application of uncovered background prediction in a low bit rate image coder", *IEEE Transactions on Communications*, vol. 38, no. 9, pp. 1578–1584, Sept. 1990.

[85] F. Lavagetto and R. Leonardi, "Block adaptive quantization of multiple frame motion field", in *Proceedings of the SPIE Conference on Visual Communications and Image Processing*, Boston, MA, Nov. 1991, vol. 1605, pp. 534–543.

[86] T. Fukuhara, K. Asai, and T. Murakami, "Very low bit-rate video coding with block partitioning and adaptive selection of two time-differential frame memories", *IEEE Transactions on Circuits and Systems for Video Technology*, vol. 7, no. 1, pp. 212–220, Feb. 1997.

[87] M. Budagavi and J.D. Gibson, "Error propagation in motion compensated video over wireless channels", in *Proceedings of the IEEE International Conference on Image Processing*, Santa Barbara, CA, Oct. 1997, pp. 89–92.

[88] M. Budagavi and J.D. Gibson, "Random lag selection in multiframe motion compensation", in *Proceedings of the International Symposium on Information Theory*, Cambridge, MA, Aug. 1998, p. 410.

[89] M. Budagavi and J.D. Gibson, "Multiframe video coding for improved performance over wireless channels", *IEEE Transactions on Image Processing*, vol. 10, no. 2, pp. 252–265, Feb. 2001.

[90] T. Wiegand, X. Zhang, and B. Girod, "Motion-compensating long-term memory prediction", in *Proceedings of the IEEE International Conference on Image Processing*, Santa Barbara, CA, USA, Oct. 1997, vol. 2, pp. 17–20.

[91] T. Wiegand, X. Zhang, and B. Girod, "Long-term memory motion-compensated prediction", *IEEE Transactions on Circuits and Systems for Video Technology*, vol. 9, no. 1, pp. 70–84, Feb. 1999.

[92] T. Wiegand, *Multi-Frame Motion-Compensated Prediction for Video Transmission*, PhD thesis, University of Erlangen-Nuremberg, Erlangen, Germany, July 2000.

[93] T. Wiegand and B. Girod, *Multi-Frame Motion-Compensated Prediction for Video Transmission*, Kluwer Academic Publishers, 2001.

[94] M. Unser and M. Eden, "Weighted averaging of a set of noisy images for maximum signal-to-noise ratio", *IEEE Transactions on Acoustics, Speech and Signal Processing*, vol. 38, no. 5, pp. 890–895, May 1990.

[95] T. Wiegand, M. Flierl, and B. Girod, "Entropy-constrained linear vector prediction for motion-compensated video coding", in *Proceedings of the International Symposium on Information Theory*, Cambridge, MA, Aug. 1998, p. 409.

[96] B. Girod, T. Wiegand, E. Steinbach, M. Flierl, and X. Zhang, "High-order motion compensation for low bit-rate video", in *Proceedings of the European Signal Processing Conference*, Island of Rhodes, Greece, Sept. 1998, pp. 253–256.

[97] W.B. Pennebaker, J.L. Mitchell, D. Le Gall, and C. Fogg, *MPEG Video Compression Standard*, Kluwer Academic Publishers, Boston, 1996.

[98] A. Gersho, "Optimal nonlinear interpolative vector quantization", *IEEE Transactions on Communications*, vol. 38, no. 9, pp. 1285–1287, Sept. 1990.

[99] R.M. Gray, "Vector quantization", *IEEE Acoustics, Speech, and Signal Processing Magazine*, vol. 1, pp. 4–29, Apr. 1984.

[100] A. Gersho and R.M. Gray, *Vector Quantization and Signal Compression*, Kluwer Academic Press, 1992.

[101] P.A. Chou, T. Lookabaugh, and R.M. Gray, "Entropy-constrained vector quantization", *IEEE Transactions on Acoustics, Speech and Signal Processing*, vol. 37, pp. 31–42, Jan. 1989.

[102] Y. Shoham and A. Gersho, "Efficient bit allocation for an arbitrary set of quantizers", *IEEE Transactions on Acoustics, Speech and Signal Processing*, vol. 36, no. 9, pp. 1445–1453, Sept. 1988.

[103] M.J. Sabin and R.M. Gray, "Global convergence and empirical consistency of the generalized lloyd algorithm", *IEEE Transactions on Information Theory*, vol. 32, no. 2, pp. 148–155, Mar. 1986.

[104] H. Everett III, "Generalized lagrange multiplier method for solving problems of optimum allocation of resources", *Operations Research*, vol. 11, pp. 399–417, 1963.

[105] K. Ramchandran, A. Ortega, and M. Vetterli, "Bit allocation for dependent quantization with applications to multiresolution and MPEG video coders", *IEEE Transactions on Image Processing*, vol. 3, no. 5, pp. 533–545, Sept. 1994.

[106] J. Lee and B.W. Dickinson, "Joint optimization of frame type selection and bit allocation for MPEG video encoders", in *Proceedings of the IEEE International Conference on Image Processing*, Austin, TX, Nov. 1994, vol. 2, pp. 962–966.

[107] J. Ribas-Corbera and D.L. Neuhoff, "Optimal bit allocations for lossless video coders: Motion vectors vs. difference frames", in *Proceedings of the IEEE International Conference on Image Processing*, Washington, DC, Oct. 1995, pp. 180–183.

[108] G.M. Schuster and A.K. Katsaggelos, "A video compression scheme with optimal bit allocation between displacement vector field and displaced frame difference", in

Proceedings of the IEEE International Conference on Acoustics, Speech and Signal Processing, Atlanta, GA, May 1996, pp. 1966–1969.

[109] G.M. Schuster and A.K. Katsaggelos, "A video compression scheme with optimal bit allocation among segmentation, motion, and residual error", *IEEE Transactions on Image Processing*, vol. 6, no. 11, pp. 1487–1501, Nov. 1997.

[110] S.-C. Han and J.W. Woods, "Adaptive coding of moving objects for very low bit rates", *IEEE Journal on Selected Areas in Communications*, vol. 16, no. 1, pp. 56–70, Jan. 1998.

[111] Y.Y. Lee and J.W. Woods, "Motion vector quantization for video coding", *IEEE Transactions on Image Processing*, vol. 4, no. 3, pp. 378–382, Mar. 1995.

[112] L.A. Da Silva Cruz and J.W. Woods, "Adaptive motion vector vector quantization for video coding", in *Proceedings of the IEEE International Conference on Image Processing*, Vancouver, Canada, Sept. 2000, vol. 2, pp. 867–870.

[113] P. Moulin, R. Krishnamurthy, and J.W. Woods, "Multiscale modeling and estimation of motion fields for video coding", *IEEE Transactions on Image Processing*, vol. 6, no. 12, pp. 1606–1620, Dec. 1997.

[114] M.C. Chen and Jr. A.N. Willson, "Rate-distortion optimal motion estimation algorithm for video coding", in *Proceedings of the IEEE International Conference on Acoustics, Speech and Signal Processing*, Atlanta, GA, May 1996, pp. 2096–2099.

[115] M.C. Chen and Jr. A.N. Willson, "Rate-distortion optimal motion estimation algorithms for motion-compensated transform video coding", *IEEE Transactions on Circuits and Systems for Video Technology*, vol. 8, no. 2, pp. 147–158, Apr. 1998.

[116] W.C. Chung, F. Kossentini, and M.J.T. Smith, "An efficient motion estimation technique based on a rate-distortion criterion", in *Proceedings of the IEEE International Conference on Acoustics, Speech and Signal Processing*, Atlanta, GA, May 1996, pp. 1926–1929.

[117] F. Kossentini, Y.-W. Lee, M.J.T. Smith, and R.K. Ward, "Predictive RD optimized motion estimation for very low bit-rate video coding", *IEEE Journal on Selected Areas in Communications*, vol. 15, no. 9, pp. 1752–1763, Dec. 1997.

[118] F. Kossentini, W.C. Chung, and M.J.T. Smith, "Rate-distortion-constrained subband video coding", *IEEE Transactions on Image Processing*, vol. 8, no. 2, pp. 145–154, Feb. 1999.

[119] M. Lightstone, *Efficient Variable Rate Image and Video Coding in an Entropy-Constrained Framework*, PhD thesis, University of California Santa Barbara, Aug. 1995.

[120] J. Besag, "On the statistical analysis of dirty pictures", *J. Roy. Statist. Soc. B*, vol. 48, no. 3, pp. 259–302, 1986.

[121] G.J. Sullivan and T. Wiegand, "Rate-distortion optimization for video compression", *IEEE Signal Processing Magazine*, vol. 15, pp. 74–90, Nov. 1998.

[122] B. Girod, "Rate-constrained motion estimation", in *Proceedings of the SPIE Conference on Visual Communications and Image Processing*, Chicago, Sept. 1994, pp. 1026–1034.

[123] T. Wiegand and B. Girod, "Lagrange multiplier selection in hybrid video coder control", in *Proceedings of the IEEE International Conference on Image Processing*, Thessaloniki, Greece, Oct. 2001, vol. 3, pp. 542–545.

[124] W. Li and E. Salari, "Successive elimination algorithm for motion estimation", *IEEE Transactions on Image Processing*, vol. 4, no. 1, pp. 105–107, Jan. 1995.

[125] Y. Lin and S. Tai, "Fast full-search block matching algorithm for motion-compensated video compression", *IEEE Transactions on Communications*, vol. 45, no. 5, pp. 527–531, May 1997.

[126] N. Merhav and M. Feder, "Universal prediction", *IEEE Transactions on Information Theory*, vol. 44, no. 6, pp. 2124–2147, Oct. 1998.

[127] L. Vandendorpe, L. Cuvelier, and B. Maison, "Statistical properties of prediction error images in motion compensated interlaced image coding", in *Proceedings of the IEEE International Conference on Image Processing*, Washington, DC, Oct. 1995, pp. 192–195.

[128] L. Cuvelier and L. Vandendorpe, "Motion estimation and compensation for CIF/SIF video sequences", in *Proceedings of the IEEE Digital Signal Processing Workshop*, Loen, Norway, Sept. 1996, pp. 21–24.

[129] L. Cuvelier and L. Vandendorpe, "Coding of interlaced or progressive video sources: A theoretical analysis", in *Proceedings of the IEEE International Conference on Image Processing*, Lausanne, Switzerland, Sept. 1996, pp. 949–952.

[130] X. Li and C. Gonzales, "A locally quadratic model of the motion estimation error criterion function and its application to subpixel interpolations", *IEEE Transactions on Circuits and Systems for Video Technology*, vol. 6, no. 1, pp. 118–122, Feb. 1996.

[131] J. Ribas-Corbera and D. L. Neuhoff, "Optimizing motion-vector accuracy in block-based video coding", *IEEE Transactions on Circuits and Systems for Video Technology*, vol. 11, no. 4, pp. 497–511, Apr. 2001.

[132] O.G. Guleryuz and M.T. Orchard, "Rate-distortion based temporal filtering for video compression", in *Proceedings of the Data Compression Conference*, Snowbird, UT, Mar. 1996, pp. 122–131.

[133] O.G. Guleryuz and M.T. Orchard, "On the DPCM compression of Gaussian autoregressive sequenzes", *IEEE Transactions on Information Theory*, vol. 47, no. 3, pp. 945–956, Mar. 2001.

[134] K. Pang and T. Tan, "Optimum loop filter in hybrid coders", *IEEE Transactions on Circuits and Systems for Video Technology*, vol. 4, no. 2, pp. 158–167, Apr. 1994.

[135] T. Wedi, "A time-recursive interpolation filter for motion compensated prediction considering aliasing", in *Proceedings of the IEEE International Conference on Image Processing*, Kobe, Japan, Oct. 1999, pp. 672–675.

[136] T. Wedi, "Adaptive interpolation filter for motion compensated prediction", in *Proceedings of the IEEE International Conference on Image Processing*, Rochester, NY, Sept. 2002, vol. 2, pp. 509–512.

[137] B. Girod and T. Micke, "Efficiency of motion-compensating prediction in a generalized hybrid coding scheme", in *Proceedings of the Picture Coding Symposium*, Tokyo, Japan, Apr. 1986, pp. 85–86.

[138] B. Girod, "Motion compensation: Visual aspects, accuracy, and fundamental limits", in *Motion Analysis And Image Sequence Processing*, M.I. Sezan and R.L. Lagendijk, Eds. Kluwer Academic Publishers, Boston, 1993.

[139] B. Girod, "Why b-pictures work: a theory of multi-hypothesis motion-compensated prediction", in *Proceedings of the IEEE International Conference on Image Processing*, Chicago, IL, Oct. 1998, pp. 213–217.

[140] A. Leon-Garcia, *Probability and Random Processes for Electrical Engineering*, Addison-Wesley Publishing Company, Inc., Reading, Massachusetts, 1994.

[141] N.S. Jayant and P. Noll, *Digital Coding of Waveforms*, Prentice-Hall, Inc., Englewood Cliffs, NJ, 1984.

[142] T. Berger, *Rate Distortion Theory: A Mathematical Basis for Data Compression*, Prentice-Hall, Englewood Cliffs, NJ, 1971.

[143] R.M. Mersereau, S.H. Fatemi, C.H. Richardson, and K.K. Truong, "Methods for low bit-rate video compression: Some issues and answers", in *Proceedings of the SPIE Conference on Visual Communications and Image Processing*, 1994, pp. 2–13.

[144] Y. Wang, J. Ostermann, and Y. Zhang, *Video Processing and Communications*, Prentice-Hall, Inc., Upper Saddle River, NJ, 2002.

[145] K. Ramchandran and M. Vetterli, "Best wavelet packet bases in a rate-distortion sense", *IEEE Transactions on Image Processing*, vol. 2, no. 2, pp. 160–174, Apr. 1993.

[146] P. Ramanathan, M. Flierl, and B. Girod, "Multi-hypothesis disparity-compensated light field compression", in *Proceedings of the IEEE International Conference on Image Processing*, Thessaloniki, Greece, Oct. 2001, vol. 2, pp. 101–104.

[147] T.M. Cover and J.A. Thomas, *Elements of Information Theory*, John Wiley & Sons, New York, 1991.

[148] S.G. Mallat, "A theory for multiresolution signal decomposition: the wavelet representation", *IEEE Transactions on Pattern Analysis and Machine Intelligence*, vol. 11, no. 7, pp. 674–693, July 1989.

[149] M. Vetterli and J. Kovacevic, *Wavelets and Subband Coding*, Signal Processing Series. Prentice Hall, Englewood Cliffs, NJ, 1995.

[150] D.S. Taubman and M.W. Marcellin, *JPEG2000: Image Compression Fundamentals, Standards, and Practice*, Kluwer Academic Publishers, 2001.

[151] J.M. Shapiro, "Embedded image coding using zerotrees of wavelet coefficients", *IEEE Transactions on Signal Processing*, vol. 41, no. 12, pp. 3445–3462, Dec. 1993.

[152] B. Usevitch, "Optimal bit allocation for biorthogonal wavelet coding", in *Proceedings of the Data Compression Conference*, Snowbird, UT, Mar. 1996, pp. 387–395.

[153] D. Taubman, "High performance scalable image compression with EBCOT", *IEEE Transactions on Image Processing*, vol. 9, no. 7, pp. 1158–1170, July 2000.

[154] I. Daubechies and W. Sweldens, "Factoring wavelet transforms into lifting steps", *J. Fourier Anal. Appl.*, vol. 4, pp. 247–269, 1998.

[155] W. Sweldens, "The lifting scheme: A new philosophy in biorthogonal wavelet constructions", in *Wavelet Applications in Signal and Image Processing III*. 1995, pp. 68–79, SPIE 2569.

[156] W. Sweldens, "The lifting scheme: A construction of second generation wavelets", *SIAM Journal on Mathematical Analysis*, vol. 29, no. 2, pp. 511–546, 1998.

[157] R. Calderbank, I. Daubechies, W. Sweldens, and B.-L. Yeo, "Wavelet transforms that map integers to integers", *Appl. Comput. Harmon. Anal.*, vol. 5, no. 3, pp. 332–369, July 1998.

[158] R.L. Claypoole, G. Davis, W. Sweldens, and R.G. Baraniuk, "Nonlinear wavelet transforms for image coding", in *Proceedings of the 31st IEEE Asilomar Conference on Signals, Systems, and Computers*, Pacific Grove, CA, Nov. 1997, vol. 1, pp. 662–667.

[159] R.L. Claypoole, R.G. Baraniuk, and R.D. Nowak, "Adaptive wavelet transforms via lifting", in *Proceedings of the IEEE International Conference on Acoustics, Speech and Signal Processing*, Seattle, Washington, May 1998, pp. 1513–1516.

[160] D. Marpe and H.L. Cycon, "Very low bit-rate video coding using wavelet-based techniques", *IEEE Transactions on Circuits and Systems for Video Technology*, vol. 9, no. 1, pp. 85–94, Feb. 1999.

[161] G. Karlsson and M. Vetterli, "Three dimensional sub-band coding of video", in *Proceedings of the IEEE International Conference on Acoustics, Speech and Signal Processing*, New York, NY, Apr. 1988, pp. 1100–1103.

[162] F. Bosveld, R.L. Lagendijk, and J. Biemond, "Hierarchical video coding using a spatio-temporal subband decomposition", in *Proceedings of the IEEE International Conference on Acoustics, Speech and Signal Processing*, San Francisco, CA, Mar. 1992, vol. 3, pp. 221–224.

[163] C.I. Podilchuk, N.S. Jayant, and N. Farvardin, "Three-dimensional subband coding of video", *IEEE Transactions on Image Processing*, vol. 4, no. 2, pp. 125–139, Feb. 1995.

[164] J. Xu, S. Li, Z. Xiong, and Y.-Q. Zhang, "Memory-constrained 3D wavelet transforms for video coding without boundary effects", in *Proceedings of the IEEE International Symposium on Intelligent Signal Processing and Communication Systems*, Honolulu, HI, Nov. 2000.

[165] J. Xu, Z. Xiong, S. Li, and Y.-Q. Zhang, "Memory-constrained 3D wavelet transform for video coding without boundary effects", *IEEE Transactions on Circuits and Systems for Video Technology*, vol. 12, no. 9, pp. 812–818, Sept. 2002.

[166] C. Parisot, M. Antonini, and M. Barlaud, "3D scan based wavelet transform for video coding", in *Proceedings of the IEEE Workshop on Multimedia Signal Processing*, Cannes, France, Oct. 2001, pp. 403–408.

[167] G. Minami, Z. Xiong, A. Wang, and S. Mehrotra, "3-D wavelet coding of video with arbitrary regions of support", *IEEE Transactions on Circuits and Systems for Video Technology*, vol. 11, no. 9, pp. 1063–1068, Sept. 2001.

[168] T. Kronander, "New results on 3-dimensional motion compensated subband coding", in *Proceedings of the Picture Coding Symposium*, Cambridge, MA, Mar. 1990, pp. 8.5–1.

[169] T. Akiyama, T. Takahashi, and K. Takahashi, "Adaptive three-dimensional transform coding for moving pictures", in *Proceedings of the Picture Coding Symposium*, Cambridge, MA, Mar. 1990, pp. 8.2–1 – 8.2–2.

[170] A. Wang, Z. Xiong, P.A. Chou, and S. Mehrotra, "Three-dimensional wavelet coding of video with global motion compensation", in *Proceedings of the Data Compression Conference*, Snowbird, UT, Mar. 1999, pp. 404–413.

[171] Y.-Q. Zhang and S. Zafar, "Motion-compensated wavelet transform coding for color video compression", *IEEE Transactions on Circuits and Systems for Video Technology*, vol. 2, no. 3, pp. 285–296, Sept. 1992.

[172] J.-R. Ohm and K. Rümmler, "Variable-raster multiresolution video processing with motion compensation techniques", in *Proceedings of the IEEE International Conference on Image Processing*, Santa Barbara, CA, USA, Oct. 1997, pp. 759–762.

[173] J.-R. Ohm, "Temporal domain sub-band video coding with motion compensation", in *Proceedings of the IEEE International Conference on Acoustics, Speech and Signal Processing*, San Francisco, CA, Mar. 1992, vol. 3, pp. 229–232.

[174] J.-R. Ohm, "Advanced packet-video coding based on layered VQ and SBC techniques", *IEEE Transactions on Circuits and Systems for Video Technology*, vol. 3, no. 3, pp. 208–221, June 1993.

[175] S.-C. Han and J.W. Woods, "Spatiotemporal subband/wavelet coding of video with object-based motion information", in *Proceedings of the IEEE International Conference on Image Processing*, Santa Barbara, CA, Oct. 1997, pp. 629–632.

[176] P. Chen and J.W. Woods, "Video coding for digital cinema", in *Proceedings of the IEEE International Conference on Image Processing*, Rochester, NY, Sept. 2002, vol. 1, pp. 749–752.

[177] J.W. Woods and T. Naveen, "A filter based bit allocation scheme for subband compression of HDTV", *IEEE Transactions on Image Processing*, vol. 1, no. 3, pp. 436–440, July 1992.

[178] C.-H. Chou and C.-W. Chen, "A perceptually optimized 3-D subband codec for video communication over wireless channels", *IEEE Transactions on Circuits and Systems for Video Technology*, vol. 6, no. 2, pp. 143–156, Apr. 1996.

[179] K.M. Uz, M. Vetterli, and D. LeGall, "A multiresolution approach to motion estimation and interpolation with application to coding of digital HDTV", in *Proceedings of the IEEE International Symposium on Circuits and Systems*, New Orleans, LA, May 1990, vol. 2, pp. 1298–1301.

[180] K.M. Uz, M. Vetterli, and D.J. LeGall, "Interpolative multiresolution coding of advanced television with compatible subchannels", *IEEE Transactions on Circuits and Systems for Video Technology*, vol. 1, no. 1, pp. 86–99, Mar. 1991.

[181] J.-R. Ohm, "Motion-compensated 3-d subband coding with multiresolution representation of motion parameters", in *Proceedings of the IEEE International Conference on Image Processing*, Austin, TX, Nov. 1994, vol. 3, pp. 250–254.

[182] S. Zafar, Y.-Q. Zhang, and B. Jabbari, "Multiscale video representation using multiresolution motion compensation and wavelet decomposition", *IEEE Journal on Selected Areas in Communications*, vol. 11, no. 1, pp. 24–35, Jan. 1993.

[183] D. Taubman and A. Zakhor, "Rate and resolution scalable subband coding of video", in *Proceedings of the IEEE International Conference on Acoustics, Speech and Signal Processing*, Adelaide, Australia, Apr. 1994, vol. 5, pp. 493–496.

[184] D. Taubman and A. Zakhor, "Highly scalable, low-delay video compression", in *Proceedings of the IEEE International Conference on Image Processing*, Austin, TX, Nov. 1994, pp. 740–744.

[185] D. Taubman, *Directionality and Scalability in Image and Video Compression*, PhD thesis, University of California, Berkeley, 1994.

[186] D. Taubman and A. Zakhor, "A common framework for rate and distortion based scaling of highly scalable compressed video", *IEEE Transactions on Circuits and Systems for Video Technology*, vol. 6, no. 4, pp. 329–354, Aug. 1996.

[187] B.-J. Kim and W.A. Pearlman, "An embedded wavelet video coder using three-dimensional set partitioning in hierarchical trees (SPIHT)", in *Proceedings of the Data Compression Conference*, Snowbird, UT, Mar. 1997, pp. 251–260.

[188] B.-J. Kim, Z. Xiong, and W.A. Pearlman, "Low bit-rate scalable video coding with 3-D set partitioning in hierarchical trees (3-D SPIHT)", *IEEE Transactions on Circuits and Systems for Video Technology*, vol. 10, no. 8, pp. 1374–1387, Dec. 2000.

[189] A. Said and W.A. Pearlman, "A new, fast, and efficient image codec based on set partitioning in hierarchical trees", *IEEE Transactions on Circuits and Systems for Video Technology*, vol. 6, no. 3, pp. 243–250, June 1996.

[190] T. Naveen, F. Bosveld, J.W. Woods, and R.L. Lagendijk, "Rate constrained multiresolution transmission of video", *IEEE Transactions on Circuits and Systems for Video Technology*, vol. 5, no. 3, pp. 193–206, June 1995.

[191] J.W. Woods and G. Lilienfield, "A resolution and frame-rate scalable subband/wavelet video coder", *IEEE Transactions on Circuits and Systems for Video Technology*, vol. 11, no. 9, pp. 1035–1044, Sept. 2001.

[192] J.Y. Tham, S. Ranganath, and A.A. Kassim, "Highly scalable wavelet-based video codec for very low bit-rate environment", *IEEE Transactions on Circuits and Systems for Video Technology*, vol. 8, no. 4, pp. 369–377, Aug. 1998.

[193] B. Felts and B. Pesquet-Popescu, "Efficient context modeling in scalable 3D wavelet-based video compression", in *Proceedings of the IEEE International Conference on Image Processing*, Vancouver, Canada, Sept. 2000, pp. 1004–1007.

[194] V. Bottreau, M. Benetiere, B. Felts, and B. Pesquet-Popescu, "A fully scalable 3D subband video codec", in *Proceedings of the IEEE International Conference on Image Processing*, Thessaloniki, Greece, Oct. 2001, pp. 1017–1020.

[195] J. Xu, Z. Xiong, S. Li, and Y.-Q. Zhang, "Three-dimensional embedded subband coding with optimal truncation (3D ESCOT)", *Applied and Computational Harmonic Analysis*, vol. 10, pp. 290–315, 2001.

[196] B. Pesquet-Popescu and V. Bottreau, "Three-dimensional lifting schemes for motion compensated video compression", in *Proceedings of the IEEE International Conference on Acoustics, Speech and Signal Processing*, Salt Lake City, UT, May 2001, vol. 3, pp. 1793–1796.

[197] L. Luo, J. Li, S. Li, Z. Zhuang, and Y.-Q. Zhang, "Motion-compensated lifting wavelet and its application in video coding", in *Proceedings of the IEEE International Conference on Multimedia and Expo*, Tokyo, Japan, Aug. 2001, pp. 481–484.

[198] L. Luo, Y. Wu, J. Li, and Y.-Q. Zhang, "3-D wavelet compression and progressive inverse wavelet synthesis rendering of concentric mosaic", *IEEE Transactions on Image Processing*, vol. 11, no. 7, pp. 802–816, July 2002.

[199] A. Secker and D. Taubman, "Highly scalable video compression using a lifting-based 3D wavelet transform with deformable mesh motion compensation", in *Proceedings of the IEEE International Conference on Image Processing*, Rochester, NY, Sept. 2002, vol. 3, pp. 749–752.

[200] J.-R. Ohm, "Motion-compensated wavelet lifting filters with flexible adaptation", in *Proceedings of the International Workshop on Digital Communications*, Capri, Italy, Sept. 2002, pp. 113–120.

[201] J. Viéron, C. Guillemot, and S. Pateux, "Motion compensated 2D+t wavelet analysis for low rate FGS video compression", in *Proceedings of the International Workshop on Digital Communications*, Capri, Italy, Sept. 2002, pp. 129–135.

[202] C. Parisot, M. Antonini, and M. Barlaud, "Motion-compensated scan based wavelet transform for video coding", in *Proceedings of the International Workshop on Digital Communications*, Capri, Italy, Sept. 2002, pp. 121–127.

[203] M. Flierl and B. Girod, "Multihypothesis motion estimation for video coding", in *Proceedings of the Data Compression Conference*, Snowbird, Utah, Mar. 2001, pp. 341–350.

[204] M. Flierl, T. Wiegand, and B. Girod, "Rate-constrained multihypothesis prediction for motion compensated video compression", *IEEE Transactions on Circuits and Systems for Video Technology*, vol. 12, pp. 957–969, Nov. 2002.

[205] A. Papoulis, *Probability, Random Variables, and Stochastic Processes*, McGraw-Hill, New York, 1991.

[206] M. Flierl and B. Girod, "Multihypothesis motion-compensated prediction with forward-adaptive hypothesis switching", in *Proceedings of the Picture Coding Symposium*, Seoul, Korea, Apr. 2001, pp. 195–198.

[207] M. Flierl, T. Wiegand, and B. Girod, "A video codec incorporating block-based multi-hypothesis motion-compensated prediction", in *Proceedings of the SPIE Conference on Visual Communications and Image Processing*, Perth, Australia, June 2000, vol. 4067, pp. 238–249.

[208] T. Wiegand, M. Lightstone, D. Mukherjee, T. G. Campbell, and S. K. Mitra, "Rate-distortion optimized mode selection for very low bit rate video coding and the emerging H.263 standard", *IEEE Transactions on Circuits and Systems for Video Technology*, vol. 6, no. 2, pp. 182–190, Apr. 1996.

[209] ITU-T Video Coding Experts Group, *Video Codec Test Model, Near Term, Version 10 (TMN-10), Draft 1, Q15-D65*, Apr. 1998, http:// standards. pictel. com/ ftp/ video-site/ 9804_Tam/ q15d65.doc.

[210] M. Flierl, T. Wiegand, and B. Girod, "Rate-constrained multi-hypothesis motion-compensated prediction for video coding", in *Proceedings of the IEEE International Conference on Image Processing*, Vancouver, Canada, Sept. 2000, vol. 3, pp. 150–153.

[211] M. Hannuksela, "Prediction from temporally subsequent pictures", Document Q15-K38, ITU-T Video Coding Experts Group, Aug. 2000, http:// standards. pictel. com/ ftp/ video-site/ 0008_Por/ q15k38.doc.

[212] ITU-T Video Coding Experts Group, *H.26L Test Model Long Term Number 9, TML-9*, Dec. 2001, http:// standards. pictel. com/ ftp/ video-site/ h26L/ tml9.doc.

[213] M. Flierl and B. Girod, "Further investigation of multihypothesis motion pictures", Document VCEG-M40, ITU-T Video Coding Experts Group, Apr. 2001, http:// standards. pictel. com/ ftp/ video-site/ 0104_Aus/ VCEG-M40.doc.

[214] M. Flierl and B. Girod, "Multihypothesis predition for B frames", Document VCEG-N40, ITU-T Video Coding Experts Group, Sept. 2001, http:// standards. pictel. com/ ftp/ video-site/ 0109_San/ VCEG-N40.doc.

[215] S. Kondo, S. Kadono, and M. Schlockermann, "New prediction method to improve B-picture coding efficiency", Document VCEG-O26, ITU-T Video Coding Experts Group, Dec. 2001, http:// standards. pictel. com/ ftp/ video-site/ 0112_Pat/ VCEG-O26.doc.

[216] T. Yang, K. Liang, C. Huang, and K. Huber, "Temporal scalability in H.26L", Document Q15-J45, ITU-T Video Coding Experts Group, May 2000, http:// standards. pictel. com/ ftp/ video-site/ 0005_Osa/ q15j45.doc.

[217] K. Lillevold, "B pictures in H.26L", Document Q15-I08, ITU-T Video Coding Experts Group, Oct. 1999, http:// standards. pictel. com/ ftp/ video-site/ 9910_Red/ q15i08.doc.

[218] K. Lillevold, "Improved direct mode for B pictures in TML", Document Q15-K44, ITU-T Video Coding Experts Group, Aug. 2000, http:// standards. pictel. com/ ftp/ video-site/0008_Por/ q15k44.doc.

[219] M. Flierl and B. Girod, "Generalized B pictures", in *Proceedings of the Workshop on MPEG-4*, San Jose, CA, June 2002.

[220] B. Girod and M. Flierl, "Multi-frame motion-compensated video compression for the digital set-top box", in *Proceedings of the IEEE International Conference on Image Processing*, Rochester, NY, Sept. 2002, vol. 2, pp. 1–4.

[221] M. Flierl, T. Wiegand, and B. Girod, "Multihypothesis pictures for H.26L", in *Proceedings of the IEEE International Conference on Image Processing*, Thessaloniki, Greece, Oct. 2001, vol. 3, pp. 526–529.

[222] D. Marpe, G. Blättermann, and T. Wiegand, "Adaptive codes for H.26L", Document VCEG-L13, ITU-T Video Coding Experts Group, Jan. 2001, http:// standards. pictel. com/ ftp/ video-site/ 0101_Eib/ VCEG-L13.doc.

[223] D. Marpe, G. Blättermann, G. Heising, and T. Wiegand, "Further results for CABAC entropy coding scheme", Document VCEG-M59, ITU-T Video Coding Experts Group, Apr. 2001, http:// standards. pictel. com/ ftp/ video-site/ 0104_Aus/ VCEG-M59.doc.

[224] T. Stockhammer and T. Oelbaum, "Coding results for CABAC entropy coding scheme", Document VCEG-M54, ITU-T Video Coding Experts Group, Apr. 2001, http:// standards. pictel. com/ ftp/ video-site/ 0104_Aus/ VCEG-M54.doc.

[225] B. Jeon and Y. Park, "Mode decision for B pictures in TML-5", Document VCEG-L10, ITU-T Video Coding Experts Group, Jan. 2001, http:// standards. pictel. com/ ftp/ video-site/ 0101_Eib/ VCEG-L10.doc.

[226] H. Schwarz and T. Wiegand, "An improved H.26L coder using lagrangian coder control", Document VCEG-HHI, ITU-T Video Coding Experts Group, May 2001, http:// standards. pictel. com/ ftp/ video-site/ 0105_Por/ HHI-RDOpt.doc.

[227] M. Flierl and B. Girod, "Investigation of motion-compensated lifted wavelet transforms", in *Proceedings of the Picture Coding Symposium*, Saint-Malo, France, Apr. 2003, pp. 59–62.

[228] M. Flierl and B. Girod, "Video coding with motion compensation for groups of pictures", in *Proceedings of the IEEE International Conference on Image Processing*, Rochester, NY, Sept. 2002, vol. 1, pp. 69–72.

[229] I.N. Bronstein and K.A. Semendjajew, *Taschenbuch der Mathematik*, Verlag Harri Deutsch, Frankfurt/Main, 1991.

[230] I.N. Bronstein and K.A Semendjajew, *Ergänzende Kapitel zum Taschenbuch der Mathematik*, Verlag Harri Deutsch, Frankfurt/Main, 1991.

Index

Algorithm
  growing, 10
  Huffman, 17
  hypothesis selection (HSA), 20, 72, 102
  iterative, 8, 20
    convergence, 20, 102
  pel-recursive, 7
  pruning, 10
Artifacts
  blocking, 9
  compression, 88
Autocorrelation function, 25
  exponentially decaying isotropic, 123
Base layer, 104
Bit allocation, 15, 32
Bit-rate, 14, 16, 101
  reduction, 30, 76, 97, 112
Block matching, 7
Boundary problem, 117
Codebook, 14, 17
Coding
  entropy, 97
    CABAC, 98
    UVLC, 98
  intra-frame, 5
  predictive, 66, 116
  subband, 2, 31, 110
Convolution, 27
Covariance matrix
  displacement error, 38, 133
  nonnegative definite, 38
  radial displacement error, 61
  singularities, 133
Decoder, 6
Decorrelation, 124
Deinterlacing, 7
Design, 14
  optimized window, 10

superimposed predictor, 14
Displacement
  accuracy
    absolute, 123
    relative, 123
  additive, 118, 121, 123
  error, 26, 118
    correlation, 2, 38
    covariance matrix, 38
    maximally negatively correlated, 41
    probability density function, 38, 123
    radial, 58
    variance, 57
  field, 7, 37
  operator, 118
  true, 26, 118
Distortion, 16, 101
  summed absolute difference (SAD), 101
  summed square difference (SSD), 100
Dual Prime Mode, 13
Dyadic decomposition, 110
Eigenvalue, 124, 134
Eigenvector, 124
Embedded block coding with optimized truncation
    (EBCOT), 31
Encoder, 6
  intra-frame, 122
Enhancement layer, 86, 104
Error propagation, 11
Extension
  cyclic, 112, 117
  symmetric, 117
Fading transitions, 88
Filter
  anti-aliasing, 24
  averaging, 47
  deblocking, 11